TECHNICAL REPORT

High-Technology Manufacturing and U.S. Competitiveness

CHARLES KELLEY, MARK WANG, GORDON BITKO, MICHAEL CHASE, AARON KOFNER, JULIA LOWELL, JAMES MULVENON, DAVID ORTIZ, KEVIN POLLPETER

TR-136-OSTP

March 2004

Prepared for the Office of Science and Technology Policy

RAND SCIENCE AND TECHNOLOGY

The research described in this report was conducted by the Science and Technology Policy Institute (operated by RAND from 1992 to November 2003) for the Office of Science and Technology Policy.

Library of Congress Cataloging-in-Publication Data

High-technology manufacturing and U.S. competitiveness / Charles T. Kelley ... [et al.].
    p. cm.
    "TR-136."
    Includes bibliographical references.
    ISBN 0-8330-3564-9 (pbk.)
    1. High technology industries—United States. 2. High technology industries—Government policy—United States. 3. Technology assessment—Economic aspects—United States. 4. Semiconductor industry. 5. Competition, International.
    I. Kelley, Charles T.

HC110.H53H526 2004
338.4'76'0973—dc22

                2004001259

The RAND Corporation is a nonprofit research organization providing objective analysis and effective solutions that address the challenges facing the public and private sectors around the world. RAND's publications do not necessarily reflect the opinions of its research clients and sponsors.

RAND® is a registered trademark.

Published 2004 by the RAND Corporation
1700 Main Street, P.O. Box 2138, Santa Monica, CA 90407-2138
1200 South Hayes Street, Arlington, VA 22202-5050
201 North Craig Street, Suite 202, Pittsburgh, PA 15213-1516
RAND URL: http://www.rand.org/
To order RAND documents or to obtain additional information, contact
Distribution Services: Telephone: (310) 451-7002;
Fax: (310) 451-6915; Email: order@rand.org

Preface

The President's Council of Advisors on Science and Technology (PCAST) recently submitted a report to the President of the United States on the relationship between high-technology manufacturing and the nation's long-term economic security. The impetus for the PCAST report is the concern that a larger share of high-tech manufacturing formerly performed in the United States is increasingly being done overseas, with potentially harmful consequences to the U.S. economy. The Science and Technology Policy Institute (S&TPI) was asked to provide analytic support to the PCAST panel. The project was sponsored by the Office of Science and Technology Policy (OSTP).

This report presents the material that S&TPI provided to the PCAST panel. It contains six essays on various topics related to high-tech manufacturing. While the essays focus on the overarching issue addressed by the PCAST panel, they can be read independently.

About the Office of Science and Technology Policy

OSTP was created in 1976 to provide the President with timely policy advice and to coordinate the federal investment in science and technology.

About the Science and Technology Policy Institute

Originally created by Congress in 1991 as the Critical Technologies Institute and renamed in 1998, the Science and Technology Policy Institute is a federally funded research and development center sponsored by the National Science Foundation (NSF) and managed by the RAND Corporation. S&TPI was managed by RAND from 1992 through November 30, 2003. The Institute's mission is to help improve public policy by conducting objective, independent research and analysis on policy issues that involve science and technology. To this end, the Institute

- supports the Office of Science and Technology Policy and other Executive Branch agencies, offices, and councils
- helps science and technology decisionmakers understand the likely consequences of their decisions and choose among alternative policies
- helps improve understanding in both the public and private sectors of the ways in which science and technology can better serve national objectives.

In carrying out its mission, the Institute consults broadly with representatives from private industry, institutions of higher education, and other nonprofit institutions.

Inquiries regarding the work described in this report may be directed to the address below.

Stephen Rattien
Director
RAND Science and Technology
1200 South Hayes Street
Arlington, VA 22202-5050
Tel: 703.413.1100, ext. 5219
Web: www.rand.org/scitech

The RAND Corporation Quality Assurance Process

Peer review is an integral part of all RAND research projects. Prior to publication, this document, as with all documents in the RAND technical report series, was subject to a quality assurance process to ensure that the research meets several standards, including the following: The problem is well formulated; the research approach is well designed and well executed; the data and assumptions are sound; the findings are useful and advance knowledge; the implications and recommendations follow logically from the findings and are explained thoroughly; the documentation is accurate, understandable, cogent, and temperate in tone; the research demonstrates understanding of related previous studies; and the research is relevant, objective, independent, and balanced. Peer review is conducted by research professionals who were not members of the project team.

RAND routinely reviews and refines its quality assurance process and also conducts periodic external and internal reviews of the quality of its body of work. For additional details regarding the RAND quality assurance process, visit http://www.rand.org/standards/.

Contents

Figures

Tables

Summary

The charter of the President's Council of Advisors on Science and Technology (PCAST) subcommittee on Information Technology Manufacturing and Competitiveness is to examine issues surrounding the migration of high-technology manufacturing from the United States to foreign countries. There is a concern that an increasing share of manufacturing—especially high-tech manufacturing—formerly performed in the United States is being done overseas with potentially harmful consequences to the U.S. economy. In particular, the rise of the semiconductor industry in Asia has been at the heart of a public debate: Should the U.S. government undertake steps to stem the migration of an industry that has meant so much to the U.S. economy from moving offshore?

The hypothesis that the nation's long-term economic security could be adversely affected by a migration of U.S. high-tech manufacturing to overseas locations is based in part on the belief that a high-tech industrial base provides the financial support and intellectual catalyst for innovative research and development (R&D). If that base stagnates, or in a worse case, declines, support for R&D may diminish. With a less vigorous R&D base, it is feared that the United States may not be able to maintain its leadership position in cutting-edge, high-tech industries. As a result, fewer students might pursue higher-education degrees in the science and engineering (S&E) fields because of reduced employment opportunities. A reduction in the number of scientists and engineers could lead to the development of fewer innovative products with a seemingly inevitable downward spiral of U.S. technological leadership and economic well-being.

PCAST submitted a report in late 2003 to the President of the United States on the relationship between high-technology manufacturing and the nation's long-term economic security. The Science and Technology Policy Institute (S&TPI) at RAND was asked to provide analytic support to the PCAST subcommittee. We focused on providing answers to the following questions:

- What are the current trends in U.S. high-tech manufacturing?
- Is there empirical evidence that the United States is in danger of losing its overall manufacturing capabilities due to foreign competition?
- Has the United States lost part of its high-tech manufacturing base in the past, and what lessons can we learn?
- How has U.S. industrial R&D changed over time? How does U.S. federal R&D compare with that of other industrial countries?
- What are the trends in the choices of academic disciplines? What fractions of advanced S&E degrees are awarded to foreign students?

- How are some other countries dealing with the migration of manufacturing to off-shore locations?

This report presents the data and analyses that S&TPI provided to the PCAST subcommittee. Given the interests of PCAST, we focus on the information technology (IT) sector generally and on computer hardware (including components) and semiconductor manufacturing specifically.

The State of U.S. High-Tech Manufacturing

To place high-tech manufacturing in context, it is worth examining the overall trends in manufacturing. The January 2003 U.S. Census Bureau's Annual Survey of Manufacturers reports that the number of U.S. manufacturing jobs has stayed relatively constant over the past 50 years, with production workers varying between 11 million and 15 million. However, the *percentage* of manufacturing jobs in the total U.S. workforce has been halved over the same period because the U.S. workforce has roughly doubled in that same time frame. After 1998, when there were 12.2 million production workers, there began a gradual decrease to 12.0 million in 1999 and 11.9 million in 2000. A sharp drop was reported in 2001 during the current U.S. economic slowdown to 11.2 million production workers, or an 8 percent drop from 1998. Data from the Bureau of Labor Statistics show this trend continuing in 2002 to 14.7 million manufacturing employees total and 10.3 million production workers.

Increased productivity is a major reason for the decreased manufacturing employment numbers. Also, the World Trade Organization's Information Technology Agreement (ITA) has increased U.S. exports of such items as semiconductor manufacturing equipment overseas and has also furthered global production networks for IT firms. The difficult issue—as yet unresolved—is separating what proportion of this decrease in manufacturing jobs is due to the U.S. recession, ITA implementation, increasing productivity, or the loss of jobs due to foreign competition.

Data segmenting the U.S. gross domestic product (GDP) by industry is available from the U.S. Department of Commerce's Bureau of Economic Analysis (BEA). Both manufacturing value added and overall GDP have increased in current dollars over past decades. Manufacturing's percentage of GDP has declined because GDP has grown faster than manufacturing value added. Manufacturing's percentage of GDP dropped from 18.7 percent in 1987 to 14.1 percent in 2001. In contrast, the services industry, which includes health, business, and legal services, grew over the same period, increasing from 16.7 percent to 22.1 percent. Other industries that increased from 1987 are the finance, insurance, and real-estate sectors, which combined rose from 17.5 percent to 20.6 percent.

With regard to high-tech manufacturing trends, there is more concern about such trends than with the overall manufacturing statistics, since the high-tech industries are in much greater flux. Many indicators, such as employment and value added per employee, showed signs of a decline in 1997–2000 before the economic slowdown in 2001. Causes for this decline included the maturing of high-tech industries and the increasing price sensitivities of its products. Decline in employment in itself is not necessarily an indication of a problem, since industrial production in semiconductors, computers, and communications equipment all skyrocketed during that same period. However, high-tech manufacturing value added per employee also dropped fast, suggesting that many of its products were rapidly becoming routine or lower-margin and hence more attractive targets for foreign manufacturing development. Falling prices in the IT sector have resulted in part from the increases in U.S. total multifactor productivity since 1995 as well as the globalization of the industry. These issues also factor directly into decreased value added for manufacturing and, correspondingly, decreased value added per employee.

Computer and electronics product manufacturing includes the subcategories of semiconductor, peripheral, terminal, and storage equipment manufacturing. Computer and electronics manufacturing is a major U.S. industry, ranking third in employment, third in manufacturing value added, and fourth in shipment value. Of the 15.9 million U.S. manufacturing employees,[1] only the fabricated metals and transportation industries employ more people than the computer and electronics industries. From 1997 to 2001, both production workers and all employment positions in the U.S. computer manufacturing industry dropped. The magnitude of the decrease in the computer manufacturing industry is larger than that for overall manufacturing. Whereas overall manufacturing employment dropped by 6 percent from 1997 to 2001, computer manufacturing employees decreased by 20 percent. The decrease of computer manufacturing production workers was even sharper, dropping by 35 percent over the same period.

The Federal Reserve Bank releases an industrial production index every year that measures the real output of the manufacturing, mining, and electric and gas utility industries. The latest released statistical data show that computers and semiconductor industrial production have rebounded since the slowdown in 2001. There was a fivefold increase in semiconductor production and a threefold increase in computer production from 1997 to 2003. Communications equipment production, however, peaked in 2000 and has continued to drop through 2003. The Federal Reserve Bank also reports on the capacity utilization of industries in an attempt to capture sustainable maximum output—the greatest level of output a plant can maintain within the framework of a realistic work schedule, after factoring in normal downtime and assuming sufficient availability of inputs. For high-tech industries, capacity utilization dropped sharply after the dot-com boom of the late 1990s, which suggests some overcapacity in high-tech industries. Since May 2002, computer and electronics manufacturing has made the largest percentage change of industries reported by the Federal Reserve Bank. Computer and electronics manufacturing and nonmetallic mineral products were the only industries to improve industrial production from May 2002 to May 2003.

The rise of foreign high-tech manufacturing industries is unmistakable. Prior to 2000, programmers in India and elsewhere were sought after because there were not enough U.S. programmers to patch Y2K software problems. With labor rates in China and India a fraction of those in the United States, labor-intensive industries, such as software development, continue to move overseas. Highly automated manufacturing processes for semiconductor and other high-tech components have a much smaller labor component than software development, but their production has also grown overseas, lured by foreign government incentives—notably from Taiwan and China. With some products like computer peripherals and monitor displays, Asia-based manufacturers are already the leaders. Peripherals and monitors are the lower value-added-per-employee end of the production scale. However,

---

[1] Manufacturing employees include production workers plus those in related support and management positions.

higher value-added-per-employee industries, such as semiconductor, computer, and storage devices, are increasingly moving overseas as well.

The ownership of overseas manufacturing can take different forms, from overseas operations of U.S.-owned companies, to U.S. joint ventures with foreign firms, to foreign companies that have grown through contract manufacturing relationships with U.S. and other foreign companies. Policymakers must consider that U.S. companies may be moving production facilities overseas to gain market access abroad or to find ways to provide lower-cost products to U.S. consumers. Those U.S. manufacturing activities that have remained in the United States tend to be the most advanced, complex manufacturing, typically requiring close coordination with engineering or design staff. But routine manufacturing, in which every efficiency must be vigorously pursued, is tending to locate overseas. Advances in supply chain IT and other business technologies increasingly make possible asset visibility and operational control of remote operations, further enabling overseas manufacturing for U.S. companies.

Is the High-Tech Manufacturing Glass Half Empty or Half Full?

When measured as a percentage of GDP, the decline in U.S. manufacturing since the mid-20th century has been so pronounced that it seems to support the idea that the United States is becoming a "postindustrial" society. This theory, made famous by sociologist Daniel Bell, posits that, as national economies develop, workers move out of relatively low-skill and low-value-added agricultural production into low-skill and then high-skill manufacturing and, at the highest stage, into high-skill, knowledge-based service production. Although Bell himself didn't put it this way, "low skill" and "high skill" are often translated to mean "low tech" and "high tech," where technological intensity is measured as an increasing function of R&D expenditures as a share of total expenditures.[2]

Most of the concerns about a possible U.S. economic shift away from manufacturing have focused on the overseas migration of manufacturing industries. Pointing to the growing U.S. deficit in manufactured goods trade, critics of the U.S. government's generally hands-off industrial and trade policies have argued that foreign governments have deliberately and successfully subsidized and promoted their own manufacturing industries at the expense of U.S. industry. In the long run, the argument goes, it is not just U.S. manufacturing firms and workers who will be hurt, but all Americans will suffer a decline in their standard of living if manufacturing is lost because of the nearsighted or simply nonexistent U.S. government policies. We label this the "deindustrialization-due-to-globalization" hypothesis.

Have apparent declines in total U.S. manufacturing output and employment since the late 1970s been caused by competition from foreign manufacturers? A closer examination of the data suggests another explanation. Data show that between 1977 and 2001, manufacturing output came close to doubling when measured in constant 1996 dollars.[3] These data suggest that, with respect to the absolute volume of production, foreign manufactured goods

---

[2] Technological intensity is also sometimes measured in terms of the proportion of scientists and engineers in the workforce.

[3] When calculated for the period 1977–2000 (before the sharp economic downturn of 2001), output slightly more than doubled.

have not replaced American manufactured goods on U.S. and world markets. In fact, over this period, U.S. manufacturers churned out more textiles, chemicals, automobiles, electronic equipment, etc., than they ever had before.[4]

Further, manufacturing's share of U.S. GDP in constant dollar terms over the 1977–2001 period declined only slightly compared with the steep decline when measured in current dollar terms. The difference between the current and constant dollar measures occurs largely because the prices of manufactured goods increased more slowly than the prices of many services. For example, construction; transportation and public utilities; health care; and financial, insurance, and real-estate service prices all rose faster than manufactured goods prices in that time frame.

Both the rising volume of output and the falling price of U.S. manufactured goods are consistent with an increase in the productivity of American workers in the manufacturing sector. The conceptual connection is as follows: As workers become more productive, fewer are needed to produce the same volume of output for a particular industry. The result is some combination of increased output and a reduced workforce. The workers who remain receive higher wages, but total costs of production fall. Therefore, as long as the industry is competitive, prices of outputs also fall. Real output per full-time equivalent manufacturing worker more than doubled between 1977 and 2001, from just under $41,000 to over $86,000 in constant 1996 dollars. In comparison, real output per service worker actually fell slightly, from $55,000 to $48,000. Thus, increases in manufacturing productivity offer an alternative explanation for observed declines in U.S. manufacturing employment and output as a share of GDP since the late 1970s.

The argument presented above has been confirmed by more sophisticated analyses using carefully constructed data and econometric techniques. Krugman and Lawrence (1994), for example, conclude that "competition from abroad has played a minor role in the contraction of U.S. manufacturing . . . . In fact, the shrinkage is largely the result of high productivity growth, at least as compared with the service sector." As they note, this is somewhat ironic because inadequate productivity growth is often blamed for the presumed loss of U.S. manufacturing competitiveness.

In a study prepared for the International Monetary Fund, Rowthorn and Ramaswamy (1997) found strong evidence that Bell's hypothesis is, in some ways, right. Most developed economies have experienced declines in manufacturing employment that, as a share of nominal GDP, are on the scale of those experienced by the United States. These declines in manufacturing employment appear to be features of successful economic development worldwide—not excepting the East Asian economies of Hong Kong, Taiwan, Singapore, and South Korea. Far from being a crisis, the so-called loss of manufacturing "is, in general, associated with rising living standards." Further, Rowthorn and Ramaswamy find that the deindustrialization in the United States and other developed economies has little to do with competition from low-wage economies; rather, it is associated with high productivity growth in manufacturing industries.

There is no empirical case to be made that the United States is in danger of losing its overall manufacturing capabilities due to foreign competition. Declines in U.S. manufactur-

---

[4] Of course, certain types of foreign manufactured goods may have replaced certain types of American manufactured goods, and there may well have been even more American manufactured goods on world markets in the absence of foreign competition.

ing employment can be directly attributed to increased manufacturing productivity, which has continued to grow since the late 1970s, when large U.S. trade deficits in manufactured goods first began to appear. Thus, more Americans than ever now find their employment in the service sector, and their numbers are continuing to grow. But it is not the case that American manufacturing per se is declining; to the contrary, American manufactured goods are more plentiful, and cheaper, than they have ever been before.

Further, we have not found empirical evidence that either supports or contradicts the notion that American high-tech industries should receive particular government support or protection. We note, however, that even when theoretically there may be economic justifications for supporting certain industries at the expense of others, the development and implementation of sound industrial policies may be difficult if not impossible. This is because, as argued by Krugman (1996a, p. 111) "you have to base interventionist proposals on detailed predictions about how firms will change their strategies in response to hypothetical policy changes, how these strategic moves will affect profits, wages, R&D, and so on, and finally, how all of these changes will affect the economy at large."

Lessons from the 1980s:
The Competition Between U.S. and Japanese Semiconductor Firms

The United States has faced challenges in high-tech manufacturing before. We examine one important case and whether the policy responses of the U.S. government were effective. Between 1979 and 1986, the U.S. percentage of the world market for DRAM (dynamic random access memory, the most common type of semiconductor memory) products fell from over 70 percent to under 20 percent and has never recovered. Japanese semiconductor manufacturers had mounted a sustained campaign to capture the market, which they successfully controlled until the emergence of South Korean competition in the 1990s. The Japanese introduced 256K DRAM chips before any U.S. firm did. As the Japanese gained market share with the 256K chip, U.S. firms filed complaints against the Japanese actions. In June 1985, the Semiconductor Industry Association filed a petition with the U.S. Trade Representative (USTR) under Section 301 of the Trade Act of 1974, alleging that there were barriers to U.S. entry into the Japanese market, that the barriers were a structural aspect of the Japanese market, that the Japanese government condoned them, and that Japanese government policy condoned overseas dumping. The USTR, in turn, decided to pursue an antidumping case by December 1985.

Under mounting pressure, Japan acceded to most of the USTR demands, in a deal designed to settle both the entry barriers and antidumping claims. The agreement had two basic conditions. First was that Japanese firms would cease dumping in all world markets, which was a substantial precedent in that a bilateral agreement dictated behavior in other markets. Related to this condition, Japanese firms had to develop detailed cost records in order to establish a price floor. The second condition addressed the market access issue in two ways. In the official agreement, Japan agreed to encourage foreign firms to achieve increased market share in the Japanese market, and, in a side letter, the Japanese government stated that it "understood, welcomed, and would make efforts to assist foreign companies in reaching their goal of a 20 percent market share within five years." Eventually, the foreign share of the Japanese market hit 20 percent in the fourth quarter of 1992. The year 1991 also marked the beginning of a sustained five-year industry boom that saw total sales revenue grow to nearly three times the 1990 level. The boom, along with greater R&D and product development associated with it, led the U.S. industry to recapture the worldwide semiconductor market leadership position in 1993 for the first time since 1985. The U.S. industry has since maintained that position.

It is important to note that industries evolve both structurally and technologically over time. Industries and product lines that once met the definition of "high tech" might not stay that way. For example, DRAM chips once involved extremely advanced technological processes and highly skilled, specialized labor. In the 1980s, the U.S. government failed to protect its industry against powerhouse Japanese firms; as a result, U.S. firms were forced to move on to products in which they were more competitive, such as microprocessors. Now DRAM technology is well known and DRAM chips have become commodity products produced competitively and at low cost across Southeast Asia. The many Japanese firms that continued to produce DRAMs are no longer earning rents—in fact, they are now operating at a large competitive disadvantage. The lesson for the U.S. government may be that the "loss" of certain industries to foreign competitors does not necessarily lead to adverse economic or national security outcomes. Further, in the long run, such losses may not be avoidable.

While unfair Japanese trade practices, specifically with regard to the manufacture of DRAMs, should have been challenged and were, those practices may have had only a transitory effect on the U.S. semiconductor industry. While the Japanese concentrated their efforts on producing what was essentially a commodity—i.e., DRAMS—the U.S. semiconductor industry focused on where its comparative advantage lies, namely, on the design and production of the more technologically advanced and innovative logic and microcontroller products.

Providing the Foundation for a Robust U.S. High-Tech Manufacturing Base

Past or present performance of the high-tech sector is no guarantee of future health. To get a glimpse into the future of the high-tech sector, we examined two essential enablers for the United States continuing to be a source of advanced technology: a robust R&D base and an adequate number of S&E graduates.

U.S. Research and Development Funding

The excellence of U.S. research universities, national laboratories, and technology industries depends critically on R&D funding for the development of emerging technologies and cutting-edge innovations. Industrial R&D funding has remained fairly steady over the past three decades relative to GDP and has increased recently.[5] This has occurred despite the fact that federal support for industrial R&D has fallen and is continuing to fall. In 1970, federal industrial R&D funding was slightly more than 1 percent of GDP. In contrast, by 2000, federal funding had fallen to about 0.25 percent of GDP. Industry has more than made up for the decline in federal support.

---

[5] Industrial R&D is performed by for-profit companies and paid for by those companies, the federal government, or other organizations and institutions.

Over the past decade, total federal R&D funding has remained fairly constant—both in terms of constant dollars and as a percentage of GDP. In 2003, total federal R&D was split nearly evenly between defense and nondefense activities. The major change has been the strong growth in the health R&D component, in which funding has increased about 150 percent since 1990. Over the same period, while federal R&D funding for industrial R&D purposes has declined, federal R&D funds to academic institutions has increased by 57 percent. The emphasis on health-related activities is reinforced when examining federal obligations for research. While the funding for all fields has generally increased over time, there has been a dramatic increase in funding in life science research, tripling in constant dollars from 1990 to 2002.

Of course, the federal R&D portfolio could be shifted back toward providing more support for industrial R&D either at the expense of other R&D activities or with an increase in the overall federal R&D budget. If such a change were contemplated, it would lead to posing the following questions: Would additional federal R&D investment help ensure the future of the high-tech industry? Or might it lead to a reduction in industry-sponsored R&D?

In aggregate measures, U.S. total R&D support is comparable to the support of other major industrial nations when measured as a percentage of GDP. In 1981, the United States spent, in constant dollars, nearly as much for R&D as that spent by all the other G-7 nations combined. By 2000, the U.S. share of total R&D had grown slightly relative to the total of those same countries. However, the United States spends its R&D funds differently than the other industrial nations. As a percentage, the United States spends considerably more on defense-related activities and considerably less on advancement-of-knowledge activities. However, there may be spillover effects from defense-related R&D activities to advancement of knowledge activities that this simple categorization does not capture. The best allocation of R&D funds for maintaining a healthy high-tech manufacturing base within the context of overall national priorities remains an open issue. However, the data do not support the thesis that difficulties that may have been experienced by high-tech manufacturing have led to a decline in support for R&D.

U.S. Science and Engineering Degrees

Ensuring a skilled workforce, one that can work and thrive in an increasingly technological and rapidly changing environment, is a U.S. priority. Of particular importance are university and college graduates with technical degrees, whose training may allow them to contribute directly to the latest innovations and technological industries. The published National Science Foundation (NSF) data on degrees granted extends only up to 1998–1999. From then until the present time, the global high-tech manufacturing situation has arguably changed more than during any other recent four-to-five-year period. It would be useful to extend the examination of data on degrees granted from 1998–1999 to the present time. However, such data were not available during the course of this study. Within the data limitations just described, we found that:

- Both the number of degrees for all disciplines and the number of S&E degrees grew at rates that are larger than the rate of growth in the U.S. workforce over the 1985–1998 period. Thus, if degrees granted per workforce member measures the need for graduates with technical degrees, the United States, it appears, would continue to produce an adequate supply of such people.

- The numbers of S&E bachelor's, master's, and doctorate degrees as a percentage of all degrees granted at the bachelor's, master's, and doctorate levels, respectively, remained remarkably constant from 1985 to 1998.

- The number of bachelor's degrees in computer science, mathematics, and electrical engineering—fields closely associated with the information technology area—as a percentage of all S&E bachelor's degrees declined from 1985 to 1998, but the percentage of graduate degrees in the three fields increased over the same period.

- Examining a limited sample of industrial countries, we found that the United States educates a larger percentage of foreign graduate students in the fields of computer science and mathematics than do other countries. At the bachelor's degree level, only a small fraction of degrees were awarded to foreign-born students. But at the graduate degree level, a significant portion of master's and doctorate degrees in the fields of computer science and mathematics awarded at U.S. academic institutions—63 percent and 71 percent, respectively—went to foreign students. While educating foreign-born students has benefits for the United States, these percentages seem to be undesirably high. The United States would appear to be very dependent on foreign-born graduates for providing the technological know-how in the computer science field—a key component of IT. The United States should consider policies for stimulating the interest of U.S. citizens in this and associated fields.

- The good news is that, in 1999, more than 80 percent of Asian students who were awarded doctorate degrees planned to stay and seek employment in the United States. Whether that trend will hold for IT fields, particularly in light of the global changes in the IT industry, is uncertain. The United States should consider balancing its desire to have foreign graduate students return to their countries, and the potential of rising to leadership positions there, with the benefit of having those same students stay in the United States—especially those who have attained advanced S&E degrees—to contribute to U.S. technological leadership.

The Taiwan–Mainland China Experience

Taiwan is experiencing an outflow of IT production to mainland China. In some respects, the situation that Taiwan is facing mirrors what many say the United States is experiencing, namely, the migration of high-tech manufacturing to overseas locations. Yet other aspects of the Taiwanese experience are unique and evolve around the particular security situation that Taiwan faces vis-à-vis mainland China. What steps has the Taiwan government taken to stem the flow of its high-tech manufacturing base to China? Have these steps yielded concrete results?

Taiwan plays a significant role in global IT production. Taiwanese companies produce approximately 60 percent of the world's notebook computers, 90 percent of its motherboards, 60 percent of its liquid crystal displays (LCDs), 50 percent of its computer display terminals (CDTs), 30 percent of its optical disk drives, and 25 percent of its servers. Until recently, Taiwan ranked third in the world in IT hardware production value, trailing only the United States and Japan. As more Taiwanese manufacturing has relocated to the mainland, however, China has surpassed Taiwan, becoming the third leading producer of IT hardware, while the island has fallen to fourth.

Prior to 1979, there was virtually no economic interaction between China and Taiwan. At the beginning of the 1980s, Taipei enforced a nearly complete ban on exports to the mainland and permitted only certain Chinese foods and medicines to be imported from China via Hong Kong. Despite these prohibitions, Taiwanese businessmen rushed to take advantage of increasing opportunities on the mainland, and cross-Strait trade reached nearly $1 billion by 1985. Perhaps recognizing the futility of enforcing the ban on trade and investment, the Taiwanese government in 1985 adopted a noninterference policy on indirect exports to and investment in China.

Yet the economy on Taiwan was undergoing important changes that would lead to an accelerated transfer of production to the mainland. Rising wages and the appreciation of the currency reduced the competitiveness of Taiwan's labor-intensive industries and forced these industries to find low-wage production bases, like China. In October 1989, Taiwan issued regulations sanctioning indirect trade with and investment in China. This mix of restrictions and tolerance allowed steady increases in trade between Taiwan and China. In 1978, Taiwanese exports to the mainland totaled a mere $51,000; however, by 1991, they had exceeded $4.6 billion.

In the early 1990s, Taiwanese investment began to move up the manufacturing chain. Migration of IT hardware manufacturing capacity across the Strait was propelled in large part by the requirement to lower production costs. In 1993, Taiwan companies producing PC-related products on the mainland were already benefiting from significant savings over the cost of production on the island, exceeding the savings available in Southeast Asian countries like Malaysia. By 1999, according to one estimate, about one-third of Taiwan's IT products were being manufactured in China.

The Taiwanese government has been playing catch-up in terms of its policies on cross-Strait economic relations. For the most part, Taiwan's policies have lagged behind economic trends, often by several years or more.[6] Nowhere has this tendency to fall behind the business curve been more evident than in Taipei's attempts to regulate the flow of investment from the island's companies into the emerging information technology sector on the mainland. Taiwan companies have found innumerable ways over the years to circumvent the restrictions imposed by the Taiwanese government, such as incorporating overseas or channeling funds through Hong Kong, the Cayman Islands, and the British Virgin Islands. In recent years, the pattern has continued, with Taiwan modifying regulations to accommodate—and to attempt to shape to the extent possible—emerging trends in the increasingly dynamic China-Taiwan economic relationship.

In many ways, the Taiwanese experience mirrors the situation that some fear is already happening to the United States. It is important to note how ineffective Taiwan's policies appear to be at stemming the tide of relocation of lower-end manufacturing to lower-cost locations and to China in particular. If Taiwan, with all its political and security concerns, cannot effectively curb the location of manufacturing to China, the U.S. government could face even more difficulty and, perhaps, less effectiveness.

---

[6] Two examples illustrate the reactive nature of Taipei's approach to economic integration with the mainland. First, Taipei did not lift its ban on private-sector cross-Strait exchanges until 1987. Second, the Taiwanese government waited until the early 1990s to legalize investment in the mainland by Taiwanese businesses.

Both China and Taiwan have a substantial array of incentives to attract foreign companies to locate in their respective territories.

China

In China, preferential policies offered by the central government and local authorities in various cities—along with the allure of a fast-growing domestic market, which stands out even more when compared with the sluggish global IT industry as a whole—have proven a potent combination, attracting large foreign investments that are fueling the growth of a nascent domestic semiconductor industry. Tax subsidies and the desire for market access are the principal draws for many companies planning to invest in semiconductor manufacturing facilities in China. The most controversial incentive is a value-added tax (VAT) rebate on chips made in China that is offered by the central government. Foreign-made chips are subject to a 17 percent VAT, while chips produced in China receive an 11 percent rebate, effectively lowering the VAT on domestically produced chips to 6 percent. The Chinese government has also offered free land use and other incentives such as tax holidays and reductions to companies building advanced semiconductor manufacturing facilities. For example, under the "2+3" incentive plan, which applies to integrated circuit manufacturers and software firms, the central government offers a two-year exemption from corporate taxes followed by a 50 percent reduction for the next three years. Local governments in China are offering their own incentives, frequently attempting to one-up each other in a fierce competition to attract foreign investment in high-tech industries. Nowhere is this regional competition more intense than in the semiconductor industry.

Taiwan

Taiwan essentially established the foundry industry, which today is dominated by two of the island's leading high-tech companies, the Taiwan Semiconductor Manufacturing Corporation (TSMC) and the United Microelectronics Corporation (UMC). In 2002, the two foundry giants combined to account for more than 70 percent of the global market for made-to-order chips. Both TSMC and UMC were spun off from a government-funded high-tech research institute. When TSMC built its first facility in 1986, the government of Taiwan contributed nearly half of the initial $200 million investment. In addition, the government reportedly offers substantial tax incentives to TSMC and UMC. Incentives designed to encourage foreign companies to establish regional headquarters and R&D centers in Taiwan include two years of free rent on land in designated industrial districts, followed by another four years of reduced rental rates, and a variety of corporate income tax breaks.

Conclusions

We revisit the questions posed earlier and summarize our conclusions:

- What are the current trends in U.S. high-tech manufacturing?
  - U.S. high-tech exports still lead the world by a large margin, and U.S. high-tech companies are expected to maintain leading market shares for some time. The latest 2003 data from the Federal Reserve Bank show that industrial production of semiconductors and computers has rebounded since the economic slowdown in

2001. However, industrial production of communications equipment has continued to drop through 2003.

- U.S. manufacturing activities that have remained in the United States tend to be the most advanced and complex manufacturing, typically requiring close coordination with engineering and design staff. But more routine manufacturing, in which every efficiency must be pursued, tends to locate overseas for economic advantages.
- However, challenges remain. For example, there has been a decrease in computer manufacturing employment that actually began several years prior to the dot-com crash. In fact, the data indicate that the United States is not facing a manufacturing crisis but rather an employment problem. While there are some linkages between the current state of U.S. manufacturing and the current unemployment situation, U.S. government policies may be better focused on employment issues, not assuming that the solution necessarily lies with the manufacturing sector.

• Is there empirical evidence that the United States is in danger of losing its overall manufacturing capabilities due to foreign competition?

- There is no empirical case to be made for this. Declines in U.S. manufacturing employment can be directly attributed to increased manufacturing productivity. The key problem is not the U.S. capability in high-tech manufacturing but rather the employment issues that have resulted from strong productivity growth.
- More Americans than ever now find their employment in the service sector, and their numbers are continuing to grow. But it is not the case that American manufacturing per se is declining. To the contrary, American manufactured goods are more plentiful, and cheaper, than they have ever been before.
- We have not found empirical evidence that either supports or contradicts the notion that American high-tech industries should receive particular government support or protection.

• Has the United States lost part of its high-tech manufacturing base in the past, and what lessons can we learn?

- Japanese semiconductor manufacturers mounted a sustained campaign to capture the DRAM market. Between 1979 and 1986 the U.S. percentage of the world market fell from over 70 percent to under 20 percent and has never recovered.
- Part of the Japanese success was due to unfair trade practices. Through a series of U.S.-Japan agreements, the United States was able to stop the dumping of Japanese semiconductors on the world market and to open the Japanese market to U.S. semiconductor firms.
- In the final analysis, Japanese practices had only a transitory effect on the U.S. semiconductor industry. While the Japanese concentrated their efforts on producing what was essentially a commodity—i.e., DRAMs—the U.S. semiconductor industry focused on where its comparative advantage lay, namely, on the design and production of more technologically advanced microprocessors and other innovative semiconductor-related products.

• How has U.S. industrial R&D changed over time? How does U.S. federal R&D compare with that of other industrial countries?

- There is no evidence that there has been a decline in industrial R&D funded or in federal funding of R&D. While federal funding of industrial R&D has declined, the shortfall has been more than made up by industry.

– Over the past decade, federal R&D funding has remained nearly the same in constant dollars, while federal R&D support to academic institutions has increased by 57 percent.
– There has been a shift in the ways federal R&D funds have been allocated over the past decade with increased emphasis on health and life sciences.
• What are the trends in the choices of academic disciplines? What fractions of advanced S&E degrees are awarded to foreign students?
—The numbers of college graduates are more than keeping pace with the growth in the U.S. workforce.
—There is no evidence that interest in attaining S&E degrees, compared with interest in non-S&E degrees, is waning.
—There is a potential problem with a large percentage of advanced IT degrees being earned by foreign students.
• How are some other countries dealing with the migration of manufacturing to offshore locations?
—Our analysis focused on investment flows between Taiwan and mainland China. Taiwan, with all its political and security concerns, has not been able to curb effectively the location of manufacturing to China.
—The U.S. government could face even more difficulty and perhaps less effectiveness if it attempts to influence investment flows.

Elements for Improving U.S. High-Technology Manufacturing

Let us set aside the debate over whether foreign economies and industries are luring away and trouncing American high-tech firms or if U.S. high-tech firms are thriving both overseas and domestically. While the debate is important, and how it is resolved will have implications for the health of the U.S. high-tech industry and for U.S. government policy, there are a number of steps that, if taken, may provide for the foundations of a consistent industrial strategy. The following strategic steps may be taken regardless of how high technology or any another manufacturing industry faces foreign competition.

The summary and findings above highlight three principal elements for improving U.S. high-tech manufacturing prospects:

• Level the Playing Field: Trade Practices
– Enforce existing trade agreements by investigating allegations of trade infractions—e.g., many consider China's VAT policy discriminatory—taking action as appropriate.
– Aggressively monitor and enforce intellectual property rights.
• Level the Playing Field: Incentives
– Determine what role, if any, the federal government should play in assisting the various U.S. states in developing incentive packages to attract high-tech industry—either U.S. or foreign.
– Consider an appropriate agreement governing such practices—e.g., the Doha round of trade talks.
• Strengthen the Innovation Infrastructure Base

– Determine if the overall magnitude of federal R&D funding is at a level appropriate for a nation that desires to remain the most technologically advanced.

– Reexamine the allocation of federal R&D funding in terms of both the types of organizations that are performing R&D and the disciplines that are being funded.

– Strengthen high school science programs so that more students will become interested in the sciences, and encourage high school students to go into S&E programs at the college level.

– Encourage S&E college graduates to continue their higher education by earning graduate degrees in S&E fields.

– Make it easier for foreign students who earn advanced S&E degrees in the United States to stay and contribute to U.S. economic growth.

Acknowledgments

This project was conducted by an interdisciplinary team of researchers at S&TPI. It would not have been possible without the support and assistance from many individuals and organizations.

RAND provided analytic support to OSTP and PCAST. The guidance received was invaluable. The chair of this PCAST panel, George Scalise, met with members of the RAND team on several occasions. He provided valuable insights on the high-technology industry and suggestions on an earlier draft of this report. From OSTP, we thank Stan Sokul for inviting the RAND team to attend and learn from a series of relevant presentations and teleconferences as well as for his availability to discuss our research in progress.

The authors especially thank Professor Matthew Slaughter of the Tuck School of Business at Dartmouth College and RAND colleague David Aaron for their insightful reviews of an earlier version of this document.

Within RAND, the authors thank Darlette Gayle, Sharon Drummond, Nora Wolverton, and Lisa Sheldone for administrative and other support. Gail Kouril contributed to the literature search, and Phillip Wirtz edited the final document.

The authors remain responsible for any errors or omissions.

Abbreviations

AFL-CIO	American Federation of Labor–Congress of Industrial Organizations
AMD	Advanced Micro Devices
ASEAN	Association of Southeast Asian Nations
ASIC	application-specific integrated circuit
ASM	Annual Survey of Manufacturers
BEA	Bureau of Economic Analysis
BLS	Bureau of Labor Statistics
BVI	British Virgin Islands
CCL	Commerce Control List
CDT	computer display terminal
CEO	chief executive officer
CFIUS	Committee for Foreign Investment in the United States
CMOS	complementary metal-oxide semiconductor
CRT	cathode-ray tube
CSPP	Computer Systems Policy Project
DCTL	direct-coupled transistor logic
DEC	Digital Equipment Corporation
DoD	Department of Defense
DOE	Department of Energy
DOI	Department of the Interior
DPP	Democratic Progressive Party [Taiwan]
DRAM	dynamic random access memory
DTL	diode transistor logic

EDAC	Economic Development Advisory Conference
EIAJ	Electronic Industries Association of Japan
EPA	Environmental Protection Agency
EPROM	erasable-programmable read-only-memory
FDI	foreign direct investment
FMV	fair market value
FTE	full-time equivalent
FY	fiscal year
G-7	Group of 7
G-8	Group of 8
GDP	gross domestic product
HDD	hard disk drive
HHS	Department of Health and Human Services
IC	integrated circuit
IPO	initial public offering
IPR	intellectual property rights
IRS	increasing returns to scale
IT	information technology
ITA	Information Technology Agreement
K	kilobit
LCD	liquid crystal display
LY	Legislative Yuan [Taiwan]
MAC	Mainland Affairs Council
MIC	Market Intelligence Center
MITI	Ministry of International Trade and Industry [Japan]
MOEA	Ministry of Economic Affairs [Taiwan]
MOS	metal-oxide semiconductor
NAFTA	North American Free Trade Agreement
NAICS	North American Industrial Classification System
NAM	National Association of Manufacturers
NASA	National Aeronautics and Space Administration
NCRA	National Cooperative Research Act

NIH	National Institutes of Health
NRC	National Research Council
NSB	National Science Board
NSC	National Science Council [Taiwan]
NSF	National Science Foundation
NTT	Nippon Telephone and Telegraph
OECD	Organisation for Economic Co-operation and Development
OSTP	Office of Science and Technology Policy
OTA	Office of Technology Assessment
PC	personal computer
PCAST	President's Council of Advisors on Science and Technology
PDA	personal digital assistant
POS	point of sale
PPM	parts per million
R&D	research and development
RCTL	resistor capacitor transistor logic
S&E	science and engineering
S&TPI	Science and Technology Policy Institute
SBS	social and behavioral sciences
SEMATECH	semiconductor manufacturing technology (program)
SIA	Semiconductor Industry Association
SIC	Standard Industrial Classification
SMIC	Semiconductor Manufacturing International Corporation
SRAM	standard random access memory
TI	Texas Instruments
TSMC	Taiwan Semiconductor Manufacturing Corporation
TSU	Taiwan Solidarity Union
TTL	transistor-transistor logic
UMC	United Microelectronics Corporation
USDA	United States Department of Agriculture
USML	United States Munitions List
USTR	U.S. Trade Representative

VAT	value-added tax
VER	voluntary export restriction
VLSI	very large scale integration
WTO	World Trade Organization
Y2K	year 2000

CHAPTER ONE

Introduction

The charter of the President's Council of Advisors on Science and Technology (PCAST) subcommittee on Information Technology Manufacturing and Competitiveness is to examine issues surrounding the migration of high-technology manufacturing from the United States to foreign countries. There is a concern that a larger share of manufacturing—especially high-tech manufacturing—formerly performed in the United States is increasingly being done overseas, with potentially harmful consequences to the U.S. economy. In particular, the rise of the semiconductor industry in Asia has been at the heart of a public debate and raises the question: What actions should the U.S. government undertake to stem the migration of an industry that has meant so much to its economy from moving offshore?

The hypothesis that the nation's long-term economic security could be adversely affected by a migration of U.S. high-tech manufacturing to overseas locations is founded in part on the belief that a high-tech industrial base provides the financial support and intellectual catalyst for innovative research and development (R&D). If the industrial base stagnates, or in a worse case, declines, support for R&D may diminish. With a less vigorous R&D base, it is feared that the United States may not be able to maintain its leadership position in cutting-edge, high-tech industries. As a result, fewer students might pursue higher-education degrees in the science and engineering (S&E) fields because of reduced employment opportunities. A reduction in the numbers of scientists and engineers could lead to the development of fewer innovative products, with a seemingly inevitable downward spiral of U.S. technological leadership and economic well-being.

PCAST submitted a report in late 2003 to the President of the United States on the relationship between high-tech manufacturing and the nation's long-term economic security. The Science and Technology Policy Institute (S&TPI) at the RAND Corporation was asked to provide analytic support to the PCAST subcommittee. Our support focused on providing answers to the following questions:

- What are the current trends in U.S. high-tech manufacturing?
- Is there empirical evidence that the United States is in danger of losing its overall manufacturing capabilities due to foreign competition?
- Has the United States lost part of its high-tech manufacturing base in the past, and what lessons can we learn?
- How has U.S. industrial R&D changed over time? How does U.S. federal R&D compare with that of other industrial countries?
- What are the trends in the choices of academic disciplines? What fractions of advanced S&E degrees are awarded to foreign students?

- How are some other countries dealing with the migration of manufacturing to off-shore locations?

This report presents the data and analyses that S&TPI provided to the PCAST subcommittee. Given the interests of PCAST, the report focuses on the information technology sector generally and on computer hardware (including components) and semiconductor manufacturing specifically.

How This Document Is Organized

The remainder of this document is organized into six chapters. Each chapter deals with a discrete aspect of the overall issue being investigated by the PCAST panel. The chapters can be read independently:

- Chapter Two presents an overview of the theoretical and empirical literature pertaining to trends in U.S. traditional and high-tech manufacturing. The chapter describes an analytic framework for understanding when and under what conditions U.S. policymakers should worry about certain trends in U.S. manufacturing and posits four exemplary scenarios for understanding the implications of how an industry is "lost" to U.S. economic welfare and to national security.

The next two chapters describe historical trends of selected high-tech manufacturing.

- Chapter Three describes the structural characteristics of the Japanese and American semiconductor industries during the 1980s and how they contributed to the rise of the Japanese industry in the semiconductor memory components. Further, the chapter describes actions the U.S. government took to help the U.S. semiconductor industry recover and whether these responses were effective.
- Chapter Four presents data on recent trends of U.S. manufacturing, with an emphasis on computer and semiconductor manufacturing statistics. It also provides statistics on foreign manufacturing of high-tech products and compares them with those of U.S. manufacturing.

The next two chapters describe two necessary ingredients for a successful high-tech manufacturing infrastructure: a solid R&D base and sufficient numbers of educated scientists and engineers.

- Chapter Five discusses long-term trends of U.S. industrial R&D funding with emphasis on the changing nature of federal government R&D funding. It also compares and contrasts U.S. and foreign R&D funding.
- Chapter Six presents trends in science and technology degrees granted at U.S. higher education institutions, including trends in degrees granted to foreign students.

Finally, we examine the migration of high-tech manufacturing from a foreign perspective.

- Chapter Seven discusses cross-Strait information technology and investment flows between Taiwan and mainland China. The chapter also focuses on some of the preferential policies being pursued by Taiwan and mainland China to attract foreign investments in high-tech manufacturing industries.

CHAPTER TWO

U.S. Traditional and High-Technology Manufacturing: An Irreversible Decline? Does It Matter?

Introduction

Recent studies commissioned by the AFL-CIO, the National Association of Manufacturers (NAM), and others have suggested that American manufacturing in the early 2000s is in crisis. The AFL-CIO, for example, argues that "the United States is losing a large share of its capacity to produce material goods" (AFL-CIO, 2003, p. 1). Robert Scott of the Economic Policy Institute calculates that large U.S. manufacturing trade deficits between 1994 and 2000 caused the net loss of 1.9 million manufacturing jobs (Scott, 2001, p. 6). The NAM study concludes that, if the U.S. manufacturing base continues to shrink at the present rate, innovations in manufacturing will shift to other global centers. If this happens, the authors of the study argue, a decline in U.S. living standards would be virtually assured (Joel Popkin and Company, 2003).

A second and related set of analyses points to the problems now facing U.S. high-tech manufacturing industries—specifically, semiconductors. For example, the National Research Council (NRC) of the National Academies of Sciences recently released a report outlining the semiconductor industry's importance to the U.S. economy and suggesting the strong need for government policies to support the industry (NRC, 2003). A white paper issued by Senator Joseph Lieberman (D-Conn.), a ranking member of the U.S. Senate's Armed Services Committee, emphasizes the severe national security implications of "the loss to the U.S. economy of the high-end semiconductor manufacturing sector [and] the potential subsequent loss of the semiconductor research and design sectors" (Lieberman, 2003, p. 1). Both analyses cite the work of economist Dale Jorgenson, who has argued persuasively that the development and deployment of semiconductor technology were largely responsible for the surge in American economic growth in the late 1990s (Jorgenson, 2001).

This chapter provides a brief overview of the theoretical and empirical literatures pertaining to trends in U.S. traditional and high-tech manufacturing. Specifically, it summarizes the findings to date with respect to the following questions:

- Is the United States in danger of losing its manufacturing base and the high-wage jobs that go with it?
- How might the loss of a U.S. high-tech industry such as microelectronics threaten the U.S. economy and national security?

The review presented is by no means complete; rather, it is intended to provide a conceptual and (to a lesser extent) empirical framework for understanding when and why U.S. policymakers should worry about certain trends in U.S. manufacturing. In particular, its goal is to distinguish among serious and less serious concerns regarding the loss of the U.S. traditional and high-tech manufacturing base.

This chapter is organized as follows. The section immediately below presents arguments for and against a theory of economic development in which economies based on manufacturing naturally give way to economies based on services. Evidence from the economics literature and some data are presented that support a particular version of the theory. The next section focuses on the arguments for and against government support of particular industries and discusses economic and national security implications of a "loss" of U.S. high-tech industries under alternative scenarios. The final section presents conclusions.

A Postindustrial Society?

The Bell Hypothesis

Data collected by the U.S. Department of Commerce seem at first to support the conclusion that U.S. manufacturing is in long-term decline.[1] As shown in Figure 2.1, value added by U.S. manufacturing industries accounted for roughly 27 percent of gross domestic product (GDP) in 1947.[2] By 2001, manufacturing's share of U.S. output had shrunk to just over 14 percent. Over the past 20 years, manufacturing employment has also fallen, from a high of over 30 percent of total full-time equivalent (FTE) workers in 1948 to just fewer than 15 percent in 2001 (see Figure 2.2). [3]

In fact, when measured as a percentage of GDP, the decline in U.S. manufacturing since the mid-20th century has been so pronounced that it seems to support the idea that the United States is becoming a "postindustrial" society. This theory, made famous by the sociologist Daniel Bell, posits that as national economies develop, workers move out of relatively low-skill and low-value-added agricultural production into low-skill and then high-skill manufacturing and, at the highest stage, into high-skill, knowledge-based service production (Bell, 1973). Although Bell himself didn't put it this way, "low skill" and "high skill" are often translated to mean "low tech" and "high tech," in which technological intensity is measured as an increasing function of R&D expenditures as a share of total expenditures (see Figure 2.3).[4]

---

[1] The Annual Survey of Manufacturers (ASM) is administered by the U.S. Census Bureau, which defines manufacturing as the mechanical, physical, or chemical transformation of materials, substances, or components into new products. The assembling of component parts of manufactured products is also considered manufacturing, except in cases in which the activity is appropriately classified as construction. The ASM surveys a representative sample of manufacturing establishments with one or more paid employees; very small firms and recent start-ups are not represented.

[2] The data presented in this section are based on the 1987 Standard Industrial Classification (SIC) system. GDP-by-industry data based on the new North American Industrial Classification System (NAICS) are scheduled for release in 2004.

[3] The data include both production and nonproduction workers.

[4] Technological intensity is also sometimes measured in terms of the proportion of scientists and engineers in the workforce. See Tyson (1992, p. 18).

Figure 2.1
Long-Term Trends in U.S. Manufacturing Output, 1947–2001 (Percentage of Gross Domestic Product)

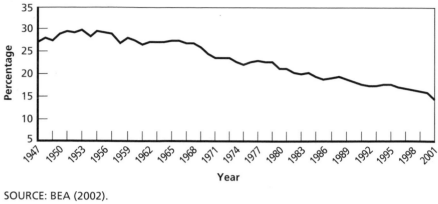

SOURCE: BEA (2002).
RAND*TR136-2.1*

Figure 2.2
Long-Term Trends in U.S. Manufacturing Employment, 1948–2001 (Percentage of Total Full-Time Equivalent Workers)

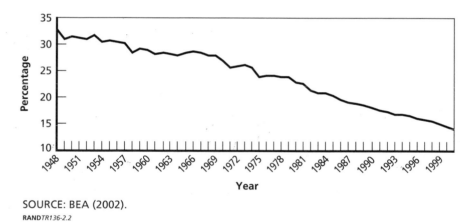

SOURCE: BEA (2002).
RAND*TR136-2.2*

Figure 2.3
Bell's Hypothesized Stages of National Economic Development

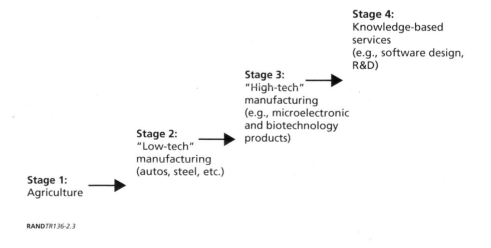

RAND*TR136-2.3*

Economists recognize Bell's hypothesis as a particular formulation of the theory of comparative advantage, which suggests that countries tend to export those products (including services) that they can produce more cheaply than can other countries.[5] As goods and services become more knowledge intensive, countries with more highly educated and highly skilled workforces can produce them most cheaply and therefore have a comparative advantage in their production. Over time, these advanced-country economies will devote a growing share of production to high-tech manufactures and knowledge-based services while importing an increasing proportion of the agricultural goods and "low-tech" manufactured goods they consume from the developing world. On net, both advanced and developing countries will benefit from such specialization. However, without some form of redistribution mechanism, workers, managers, and shareholders in the declining sectors of the advanced economy would be (at least temporarily) worse off.

Critics of Bell—or, more accurately, critics of those who argue against government support of the manufacturing sector on the basis of comparative advantage—have questioned both the inevitability and the desirability of any shrinkage of traditional manufacturing industries such as automobiles and steel. Stephen Cohen and John Zysman, for example, made the following observations about what they saw as a dangerous tendency to accept the decline of traditional manufacturing industries—and of all manufacturing production—in the United States as natural and inevitable:

- Traditional low-tech manufacturing is essential to high-tech manufacturing because most high-tech products are producer goods (i.e., inputs into traditional production), not consumer goods.
- Manufacturing production is essential to the high-tech sector because profits from production finance most R&D.
- There is no natural progression from high-tech manufacturing to knowledge-based services because knowledge-based service jobs (i.e., high-skill, high-wage jobs) are complements, not substitutes, for "good" manufacturing jobs.
- Knowledge-based services, such as R&D, are dependent on manufacturing because of features like learning by doing: "Unless R&D is tightly tied to the manufacturing of the product, [it] will fall behind the cutting edge of incremental innovation."[6]

Arguing that manufacturing is critical to the wealth and power of the United States, Cohen and Zysman and others have proposed a set of generic as well as sector-specific domestic and trade policies designed to keep traditional as well as high-tech American manufacturers healthy—and at home (Bluestone and Harrison, 1982; OTA, 1988, 1990, 1991; Kuttner, 1991; AFL-CIO, 2003).

Why—and to What Extent—Has U.S. Manufacturing Declined?

Most of the concerns about a possible U.S. economic shift away from manufacturing have focused on the overseas migration of manufacturing industries. Pointing to the growing U.S. deficit in manufactured goods trade (Figure 2.4), critics of the U.S. government's generally

---

[5] See Ricardo (1817) for the classic exposition of the theory of comparative advantage.

[6] Cohen and Zysman (1987, p. 8). The authors also point out that agriculture, and many supposedly low-tech manufacturing industries, rely far more on R&D than simple statistics suggest.

hands-off industrial and trade policies have argued that foreign governments have deliberately and successfully subsidized and promoted their own manufacturing industries at the expense of U.S. industry (OTA, 1988, 1990, 1991; Kuttner, 1991; Tyson, 1992; Thurow, 1994). In the long run, the argument goes, it is not just U.S. manufacturing firms and workers who will be hurt; all Americans will suffer a decline in their standards of living if manufacturing is lost due to nearsighted or simply nonexistent U.S. government policies. We label this the "deindustrialization-due-to-globalization" hypothesis.

Although the distinction is not always clearly made, the deindustrialization-due-to-globalization hypothesis suggests that American firms and workers are being squeezed internationally from two different directions. At the "traditional," or low-tech, end of the scale is the competition from firms in low-wage, low-regulation developing economies that is forcing U.S. companies to either relocate their manufacturing operations abroad or close shop altogether.[7] In industries in which R&D is at a premium, it has been claimed that Americans are losing ground to better-educated and more innovative Europeans and East Asians.[8]

The Office of Technology Assessment (OTA, 1988, p. 5) states, for example, "There are signs that the United States is losing its once substantial edge in technology, a crucial factor in competitiveness for an advanced, high-wage nation." OTA and others have argued that:

- U.S. spending on civilian R&D is too low relative to such countries as Japan and Germany.
- The U.S. education system is failing to turn out sufficient numbers of well-trained students, especially in science and engineering.

Figure 2.4
U.S. Manufactured Goods Trade Balance, 1976–2002

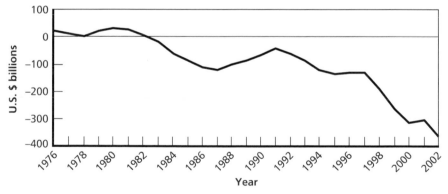

SOURCE: International Trade Administration, U.S. Department of Commerce.
RAND*TR136-2.4*

---

[7] The industries at greatest risk from this type of competition are labor intensive, such as toy and apparel manufacturing.

[8] In the past, East Asian competition in high-tech areas has come primarily from Japan, and to a lesser extent, South Korea and Taiwan. More recently, however, China has been identified as an emerging high-tech rival. We touch on this debatable point in our conclusion.

- American manufacturers have fallen behind in the practical application of technology.

This combination of factors has created a scenario in which American companies are trying to compete on world markets using relatively unskilled (but highly paid) American workers, old ideas, and old machines. As a result, the story goes, advanced economies such as Japan's and the European Union's are beating the United States in the high-tech arena, forcing American workers into competition with the low-skill, low-wage masses of the developing world. Therefore, instead of moving up Bell's development ladder toward the "knowledge-based services" stage, the United States is at risk of becoming a nation of hamburger flippers.

But have apparent declines in total U.S. manufacturing output and employment since the late 1970s been caused by competition from foreign manufacturers? A closer examination of the data suggests another plausible explanation.

As shown in Figure 2.5, between 1977 and 2001, manufacturing output came close to doubling when measured in constant 1996 dollars.[9] These data suggest that, with respect to the absolute volume of production, foreign manufactured goods have not replaced American manufactured goods on U.S. and world markets. In fact, over this period, U.S. manufacturers churned out more textiles, chemicals, automobiles, electronic equipment, etc., than they ever had before.[10]

Further, as shown in Figure 2.6, manufacturing's share of U.S. GDP in constant dollar terms over the 1977–2001 period declined only slightly compared with the steep decline when measured in current dollar terms. The difference between the current and constant dollar measures occurs largely because the prices of manufactured goods increased more

Figure 2.5
Real Manufacturing Output, 1977–2001

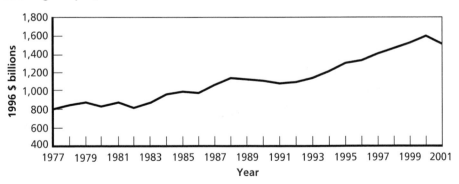

SOURCE: BEA (2002).
RAND*TR136-2.5*

---

[9] When calculated for the period 1977–2000 (before the sharp economic downturn of 2001), output slightly more than doubled.

[10] Of course, certain types of foreign manufactured goods may have replaced certain types of American manufactured goods, and there may well have been even more American manufactured goods on world markets in the absence of foreign competition.

slowly than the prices of many services. For example, construction; transportation and public utilities; health care; and financial, insurance, and real-estate service prices all rose faster than manufactured goods prices from 1977 to 2001 (see Figure 2.7).

Both the rising volume of output and the falling price of U.S. manufactured goods are consistent with an increase in the productivity of American workers in the manufacturing sector. The conceptual connection is as follows: As workers become more productive, fewer are needed to produce the same volume of output for a particular industry. The result is some combination of increased output and reduced workforce.[11] The workers who remain receive higher wages, but total costs of production fall. Therefore, as long as the industry remains competitive, prices of outputs will also fall.

Figure 2.6
Manufacturing Output as a Share of GDP, 1977–2001

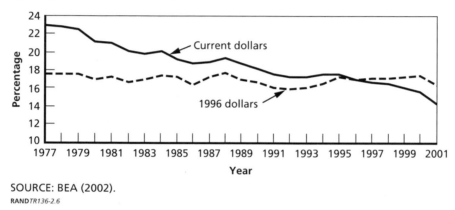

SOURCE: BEA (2002).
RAND*TR136-2.6*

Figure 2.7
Implicit Price Deflators for Four GDP Components, 1977–2001

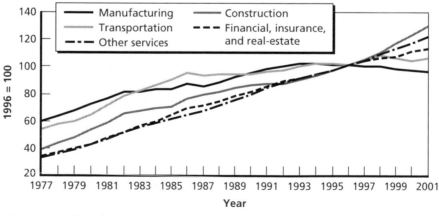

SOURCE: BEA (2002).
RAND*TR136-2.7*

---

[11] How much of the productivity improvement is realized as an increase in output, versus a decline in the workforce, depends on the nature of demand.

As shown in Figure 2.8, real output per FTE manufacturing worker more than doubled between 1977 and 2001, from just under $41,000 to just over $86,000 in constant 1996 dollars. In comparison, real output per service worker actually fell slightly, from $55,000 to $48,000.[12] An alternative measure of productivity—multifactor productivity—shows a similar trend for the U.S. manufacturing sector, increasing by more than 40 percent between 1977 and 2000.[13] Thus, increases in manufacturing productivity offer an alternative explanation for observed declines in U.S. manufacturing employment and output as a share of GDP since the late 1970s.

The simple graphical analysis presented above has been confirmed by more sophisticated analyses using carefully constructed data and econometric techniques.[14] Krugman and Lawrence (1994), for example, conclude that "competition from abroad has played a minor role in the contraction of U.S. manufacturing. . . . In fact, the shrinkage is largely the result of high productivity growth, at least as compared with the service sector." As they note, this is somewhat ironic because inadequate productivity growth is often blamed for the presumed loss of U.S. manufacturing competitiveness.

Similarly, in a study prepared for the International Monetary Fund, Rowthorn and Ramaswamy (1997) found strong international evidence that Bell's hypothesis is, in some ways, right. Most developed economies have experienced declines in manufacturing employment that, as a share of nominal GDP, are on the scale of those experienced by the United States. These measures of declines in manufacturing employment appear to be features of successful economic development worldwide—not excepting the East Asian economies of Hong Kong, Taiwan, Singapore, and South Korea (see Figures 2.9, 2.10, and 2.11).

Figure 2.8
Real Output per Full-Time Equivalent Worker, 1977–2001

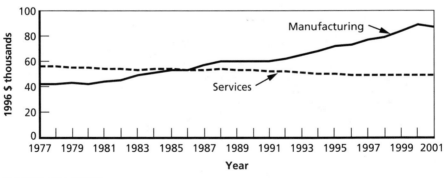

SOURCE: BEA (2002).
RAND*TR136-2.8*

---

[12] This category, represented as "other services" in Figure 2.7, includes food, lodging, amusement, recreational, legal, health, education, business, and personal services.

[13] Multifactor productivity (sometimes referred to as total factor productivity) "measures the joint influences on economic growth of technological change, efficiency improvements, returns to scale, reallocation of resources, and other factors" (U.S. Bureau of Labor Statistics, 2002).

[14] In many of these studies, productivity growth is represented by multifactor productivity. See, for example, Lawrence and Slaughter (1993).

Far from being a crisis, "loss" of manufacturing "is, in general, associated with rising living standards" (Rowthorn and Ramaswamy, 1997, p. 14). Further, Rowthorn and Ramaswamy find that the decline in manufacturing employment in the United States and other developed economies has little to do with competition from low-wage economies. Rather, as suggested above, it is associated with high productivity growth in manufacturing industries.[15]

What has happened to all those workers who, according to Scott (2001) and others, have lost their jobs in manufacturing? It turns out that U.S. manufacturing employment has actually declined very little in absolute terms, falling from a high of 21 million workers in 1979 to just over 17 million in 2001 (see Figure 2.12). [16] To offer some perspective, total U.S. employment, measured as FTE workers, grew by almost 38 million workers over the same period (BEA, 2002).[17] Certainly, U.S. civilian unemployment in the 1980s and 1990s remained on average quite low.[18]

Figure 2.9
Employment by Sector as a Share of Total Civilian Employment

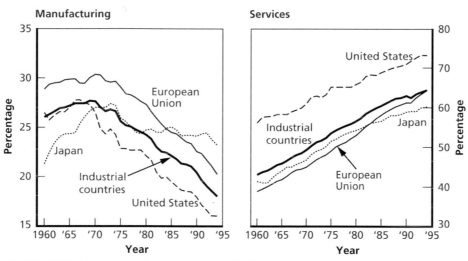

SOURCE: OECD historical statistics as presented in Rowthorn and Ramaswamy (1997, p. 8).
RAND*TR136-2.9*

---

[15] In the case of the United States and Japan, however, patterns of trade did contribute somewhat to changes in manufacturing employment. Rowthorn and Ramaswamy estimate that bilateral trade deficits with Japan contributed about 1 percentage point to the decline in U.S. manufacturing employment between 1970 and 1994, with the corresponding Japanese trade surpluses increasing manufacturing employment in Japan by about the same amount.

[16] Some of the apparent decline in U.S. manufacturing employment (and apparent increase in manufacturing productivity) may also be due to domestic outsourcing of nonmanufacturing activities (such as facilities maintenance) that had formerly been performed in-house. See, for example, Estavao and Lach (1999).

[17] BEA (2002) data are based on BEA GDP by industry; manufacturing FTE numbers are available online at www.bea.gov.

[18] U.S. unemployment as a percentage of the civilian labor force aged 16 and older averaged 6.3 percent between 1980 and 2002 (1980–2002 average unemployment rate based on data available from the U.S. Bureau of Labor Statistics, downloaded from www.bls.gov). Economists generally estimate the "natural," or nonaccelerating-inflation, rate of unemployment at between 5 percent and 7 percent (CBO, 2002).

Figure 2.10
Value Added by Sector as a Share of Total Civilian Employment

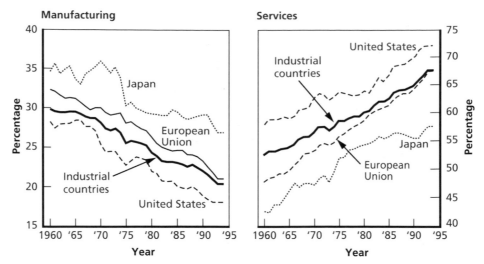

SOURCE: OECD historical statistics as presented in Rowthorn and Ramaswamy (1997, p. 8).
RAND*TR136-2.10*

Figure 2.11
Share of Manufacturing in Employment (Selected East Asian Countries)

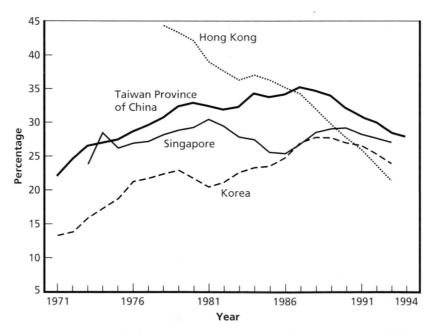

SOURCE: International Labor Organization's *Yearbook of Labor Statistics* and the
Statistical Yearbook of the Republic of China, as presented in Rowthorn and
Ramaswamy (1997, p. 16).
RAND*TR136-2.11*

Figure 2.12
U.S. Manufacturing Employment, 1948–2001 (Full-Time Equivalent Workers)

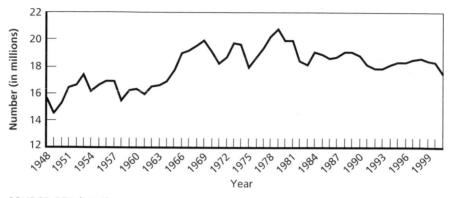

SOURCE: BEA (2002).
RAND*TR136-2.12*

However, the process of deindustrialization has not affected all industries or all regions of the country equally. For example, Kletzer (2001, p. 3) finds that import-competing job loss has been particularly concentrated "in a few large-employment industries: electrical machinery, apparel, motor vehicles, nonelectrical machinery, and blast furnaces."[19] Because several of these industries are heavily concentrated geographically, some areas of the country, such as the upper Midwest, have suffered more than others. Further, deindustrialization has been associated with a dramatic increase in the inequality of earnings between skilled and unskilled workers.[20] There is an ongoing debate over the extent to which international trade is to blame for this, but Cline (1999) attributes about one-third of the 18 percent increase in the skilled/unskilled wage differential from 1973 to 1993 to competition from imports.[21]

Loss of High-Tech Manufacturing: A Framework for Evaluation

The evidence presented above suggests there is little reason to fear a generalized loss of U.S. manufacturing capability: Although manufacturing employment has decreased dramatically as a share of total U.S. employment, American companies and workers are making more products than they ever have before—and are making them more efficiently. The evidence presented above does not, however, address the question of whether they are making the *right* things. While overseas migration of the toy industry, or even consumer electronics, may not have negative repercussions on a national scale, what about passenger aircraft? Or pharmaceuticals? Or semiconductors? Do these "high-tech" industries have characteristics that

---

[19] Kletzer (2001) also finds that relatively few displaced manufacturing workers end up flipping hamburgers. About half find new jobs in manufacturing, while just 10 percent become reemployed in the retail trade sector, which includes, among other things, McDonald's and other fast-food establishments.

[20] Economists use several measures to stratify the "skill" level of workers, including education level, work experience, and occupation. According to Lawrence and Slaughter (1993, p. 162), no matter how the skilled are distinguished from the unskilled, "the sharp rise in wage inequality between the two in the 1980s is clear."

[21] According to one of our reviewers, Cline (1999) is at the high end of such estimates. Johnson and Slaughter (2001) carefully discuss the many possible determinants of the rising inequality of earnings between skilled and unskilled workers.

make them more deserving of promotion or protection than other industries? Are foreign firms and foreign workers poaching industries that are somehow vital to the United States?

In this section, the following two primary dimensions for evaluating whether an industry should be singled out for special treatment are considered:

- U.S. national security
- U.S. national economic welfare.

With respect to these two dimensions, we explore some of the implications of the organizational separation of the production process into R&D and manufacturing production and the geographic separation of ownership from R&D and/or manufacturing production location. As shown in the scenarios developed below, these distinctions matter not only because of their implications for which policies are likely to be most effective at preserving U.S. national security and economic welfare, but also for determining whether additional policies are even necessary.

National Security Rationales for Industrial Targeting

Two of the primary national security rationales for the promotion, protection, and/or regulation of specific industries are the need to

- prevent actual or potential enemies from obtaining products, components, or technologies that could be used to harm the United States or its friends and allies
- ensure the security of supply of military-critical products, components, and technologies for the U.S. defense industrial base.

Policies designed to meet the first requirement tend to emphasize regulation, while those for the second generally involve industrial promotion and protection as well.

To control outward flows, the U.S. government strictly regulates the export or transfer to foreigners of all defense-related items, technologies, and services that have been placed on the U.S. Munitions List (USML). The U.S. State Department designates what belongs on the USML. Dual-use items and technologies—i.e., items and technologies with both military and civilian applications—are listed on the Commerce Control List (CCL), which is administered by the U.S. Department of Commerce. Recent high-level efforts by industry to liberalize both the USML and the CCL have made little progress, despite a significant measure of support from the U.S. Department of Defense (DoD).[22]

The Committee for Foreign Investment in the United States (CFIUS) and the National Industrial Security Program control foreign access to items and technologies within the United States. CFIUS, which is headed by the U.S. Treasury Department, is empowered to prohibit or dissolve all foreign acquisitions, mergers, or takeovers of U.S. firms that are deemed to threaten U.S. national security. "National security" is interpreted here quite broadly, and factors considered include "the potential effects of the transaction on U.S. technological leadership in areas affecting U.S. national security" (U.S. Department of the

---

[22] See Lorell et al. (2002) for a discussion of issues surrounding the USML, CCL, and various efforts to reform and restructure national and international export control regimes. How and to what extent to control U.S. commercial-grade microprocessors has been among the more controversial issues.

Treasury, undated). This means, for example, that a foreign-owned company seeking to invest in a U.S. company conducting leading-edge electronics or materials science R&D would likely be subject to CFIUS review.[23] The National Industrial Security Program, which is administered by DoD, prevents foreigners, including the foreign owners and foreign employees of U.S.-located companies, from accessing classified information (Lorell et al., 2002).

Ensuring the security of supply of defense-related products, components, and technologies is in some ways a more difficult problem for the U.S. government than controlling foreign access. One reason is that, for many dual-use technologies and products (such as semiconductors), military demand is only a small percentage compared with that of commercial demand. While the U.S. government, and specifically DoD, takes the health of the U.S. defense industrial base into account in its acquisition policies, it has limited leverage over the individual business decisions of U.S. firms. Security of supply becomes a particular problem when U.S. firms discontinue production of older dual-use technologies and parts and the only manufacturers of spare parts for legacy systems are located overseas.[24]

Nevertheless, in many cases, the U.S. government *does* seem to indulge its preference for American-made defense products. For example, whereas the United States accounted for 19 percent of total world imports of manufactured goods, and 22 percent of world imports of office machines and telecommunications equipment, in 1999, U.S. imports of finished conventional armaments amounted to no more than 3 percent of the world total (Lorell et al., 2002). A 2001 DoD study indicated that the value of parts, components, and materials subcontracted from foreign sources accounted for less than 2 percent of the value of total subcontracts for eight large weapon system programs (DoD, 2001, p. 17). Under current law, the minimum threshold for U.S. content on Pentagon purchases is 50 percent.[25]

In sum, there are numerous laws, regulations, and policies designed to keep defense-related U.S. products and technologies out of hostile hands and to prevent excessive reliance on overseas sources of supply. If anything, in recent years, DoD has concluded that current levels of regulation are too restrictive, thwarting its attempts to achieve greater interoperability between American forces and those of U.S. friends and allies, as well as limiting its ability to obtain the best products at the best prices (AIAA, 2001; Lorell et al., 2002; Wayne, 2003). It is important to note, however, that many of these constraints apply to items perceived to be militarily critical. For dual-use items and technologies—many of which are high tech—the constraints are not always so clear.

Economic Rationales for Industrial Targeting

The preferential targeting of individual industries usually lowers national welfare because governments are less efficient than markets at allocating resources across alternative uses. History has shown that private, broad-based financial and venture capital markets are generally better than small groups of decisionmakers at identifying promising new areas for investment

---

[23] Notification to CFIUS of proposed foreign investments is voluntary. Although compliance is generally believed to be quite high, the U.S. General Accounting Office (GAO, 2000) found at least three cases in which CFIUS should have been notified but was not.

[24] See, for example, the discussion of out-of-production parts in Lorell et al. (2000).

[25] As of August 2003, proposed congressional legislation would raise this threshold to 65 or even 70 percent. DoD and many defense and high-tech firms are opposed to the legislation (Phillips, 2003).

and that countries are better off when they specialize in areas in which they have a comparative advantage. Further, in the context of international trade, subsidization of targeted industries is often undesirable because it tends to lower the prices commanded by exports on competitive world markets. Finally, as a practical political matter, many economists oppose targeting because resources are likely to be wasted on lobbying efforts by industries seeking special treatment (Grossman, 1986; Krugman, 1996a).

There is, however, a theoretical case for the governmental protection and promotion of a particular industry if

- costs fall and product quality improves as the scale of production increases, creating first-mover advantages and strategic behavior in the industry;
- noneconomic barriers to entry (such as protectionist trade policies) create first-mover advantages and strategic behavior in the industry; and/or
- the industry creates large spillover benefits for other economic activities (Helpman and Krugman, 1985; Tyson, 1992).

Arguably, technology-intensive industries exhibit all three of these features. What, therefore, do we suffer if we lose these industries to foreign competitors?

Increasing returns to scale (IRS), steep learning curves, the dynamics of innovation, and other barriers to entry that create oligopolistic market structures generate above-normal profits, or "rents," which are distributed as returns to capital in the form of dividends and capital gains and to labor in the form of wages and benefits.[26] The share of rents that each group is able to claim depends on the relative power of workers versus shareholders and managers. One potential loss from high-tech deindustrialization, therefore, is the high rents from abroad and the corresponding increases in national wealth generated through economic multiplier effects.

High-tech deindustrialization may also reduce national economic welfare if the R&D activities of high-tech firms generate significant spillover benefits. These benefits can be placed into three broad categories: consumer surplus that is not fully captured in the price of new or improved products; growth and employment externalities associated with R&D and production facilities; and inherent contributions to public goods, such as national security and public health (Bonomo et al., 1998, p. 12). Of these, arguments favoring preferential targeting of high-tech industries tend to focus on their alleged growth and employment externalities, which are generated through the intentional and unintentional transfer of technology to other industries and activities. As discussed below, such beneficial transfers are often assumed to be a positive function of geographic proximity. In other words, those industries and activities that lie closest to R&D centers are the ones most likely to benefit from any spillovers generated.

Ownership and Location Possibilities for a Globalized High-Tech Industry

In many if not most high-tech industries in the United States, the R&D activities of firms are not only organizationally separate from manufacturing production but are also locationally separate. Research divisions have their own management within the firm and may or

---

[26] As defined by Krugman (1986, p. 12), *rent* is "payment to an input higher than what that input could earn in an alternative use."

may not be housed at the same physical location as the firm's headquarters or manufacturing assembly lines. In some industries, the separation of R&D from manufacturing has progressed even further, with R&D assets under completely separate ownership from manufacturing assets. For example, in the semiconductor industry, there has been significant growth in the number of specialized independent design houses, or "fabless firms," that rely on third-party foundries to fabricate the devices they design. According to Macher, Mowery, and Hodges (1999, p. 268), the standardization of manufacturing technologies for commercial semiconductor devices through MOS (metal-oxide semiconductor) manufacturing processes has been instrumental in giving fabless firms access to foundry capacity.

Further, as more and more countries have opened their borders to international investment, American high-tech firms have joined their more low-tech cousins in geographically segregating their activities internationally.[27] Both proximity to customers and compliance with foreign countries' local content laws have been important factors in the overseas location of manufacturing and/or final assembly of high-tech products. But the standardization of manufacturing processes that, in many industries, have made it easier to separate R&D activities from manufacturing production have also made it more attractive to choose production locations on the basis of lowest cost. In the case of hard disk drives (HDDs), for example, "a dramatic change in the locus of assembly occurred" between 1983, when just four U.S. companies had assembly operations overseas, and 1990, when Americans firms assembled two-thirds of their HDDs in Southeast Asia (McKendrick, 1999, p. 302).[28] Similar patterns of manufacturing out-migration have been found in semiconductors and other high-tech industries.[29]

Table 2.1 shows the many possible combinations of ownership and location for R&D activity, manufacturing activity, and end-product market for a globalized firm.[30] So, for example, the combination A-C represents the case in which a U.S.-owned firm conducts R&D in the United States, creating designs that it transfers or contracts out to a U.S.-owned manufacturing operation located abroad.[31] The final product is then exported back to the United States for purchase by U.S. households, firms, or government. The combination A-F represents the case in which an independent U.S.-owned R&D firm operating in the United States contracts out to a foreign-owned but U.S.-located manufacturing firm that exports to foreign markets. The combination C-G represents the case of a foreign-owned firm that conducts R&D in the United States, creating designs that it transfers or contracts out to a foreign-owned manufacturing operation located overseas. In this case, the final product becomes destined for the U.S. market.

---

[27] The concept of an "American" firm in itself becomes increasingly problematic when applied to firms with globally diversified assets. The owners and managers of such firms are often highly diversified in terms of their nationalities.

[28] Fabrication of HDD parts and components has also moved steadily into Southeast Asia. Citing Gourevitch, Bohn, and McKendrick (1997), McKendrick (1999, p. 307) states: "By 1995, more than 60 percent of global employment in the HDD industry, including upstream activities, was in Asia outside of Japan."

[29] See, for example, Grunwald and Flamm (1985). More recent evidence presented by firms at meetings of the PCAST also strongly supports the view that high-tech firms increasingly are choosing to locate their manufacturing operations overseas.

[30] More combinations are possible if the R&D and manufacturing phases are separated into finer categories. For example, many U.S. semiconductor firms have established separate "development facilities" to support the development and debugging of new process technologies and equipment (Macher, Mowery, and Hodges, 1999).

[31] A firm transfers if the R&D and manufacturing activities are conducted within an integrated firm; it contracts out if independent firms conduct R&D and manufacturing.

Table 2.1
Ownership and Location Combinations for R&D, Manufacturing Activity, and End-Product Market

Activity	Location of Activity[a]	Location of Market for Final Product	U.S. Ownership[c]	Foreign Ownership[c]
Research and development[b]	U.S.	—	A	C
	Abroad	—	B	D
Manufacturing production[b]	U.S.	U.S.	A	E
		Abroad	B	F
	Abroad	U.S.	C	G
		Abroad	D	H

[a]Location of manufacturing activity is also the location of market for R&D product.
[b]R&D and manufacturing may be conducted either by independent design houses or foundries or by divisions of integrated firms.
[c]The letters are shorthand for designating the various ownership-location combinations.

Empirically, each of these combinations would be reflected in a different set of data categories for international transactions. Combination A-C would show up in the data as U.S. direct investment abroad (the U.S.-owned manufacturing firm located overseas), as U.S. merchandise imports (sale of the final product back to the U.S. market), and probably as U.S. service exports (licensing or "export" of the U.S. design to the foreign-located manufacturer). Combination A-F would appear as foreign direct investment in the United States (the foreign-owned manufacturer located in the United States) and U.S. merchandise exports. Combination C-G would appear as foreign direct investment, U.S. service exports, and U.S. merchandise imports.

High-Tech Deindustrialization: Four Scenarios

If a high-tech industry is deserving of governmental promotion or protection, the way in which it is "lost" has implications for U.S. economic welfare or national security, or both. It matters, for example, whether U.S. firms in an industry mostly relocate overseas (in which case the ownership remains American but the workers are foreign) or shut down altogether (in which case foreign competitors take over and both owners and workers are foreign). It also is likely to matter whether it is a firm's manufacturing or R&D operations, or both, that migrate overseas. Four simple but familiar scenarios illustrate the importance of correctly characterizing the nature of U.S. high-tech deindustrialization for policy purposes.

Scenario 1: U.S. Ownership/U.S. Control Over All Phases of Production. In this scenario, combinations A-A or A-B prevail. R&D and manufacturing assets are owned by Americans, and all phases of production are carried out by American workers. The final product is either sold on the U.S. market or exported abroad. If the industry earns rents, they are split between owner-managers and workers. Any technological spillovers are retained in the United States. If the industry produces items that are militarily sensitive, any merchandise and technology exports would likely be subject to governmental controls.

Scenario 2: U.S. Ownership/Foreign Manufacturing Location. In this scenario, manufacturing moves overseas and combinations A-C or A-D prevail. If the industry is oligopolistic, American manufacturing workers lose their high-wage jobs while American owners continue to earn above-normal returns. R&D workers—mostly scientists and engineers—keep their high-wage jobs.

This scenario could be undesirable on the grounds of economic equity if the move overseas represents an attempt by owners to divert a greater share of global rents their way.[32] However, if the reason for the move is to avoid trade barriers or friction with trading partners, one could argue that no American jobs are lost because it would not have been possible to export from an American manufacturing location.[33] If the industry in question generates technological spillovers, then presumably they are retained because the R&D activity stays in the United States. However, the nature of technological spillovers is not well understood. It seems likely that spillovers involving innovations to the manufacturing process rather than to the product design would accrue to the foreign production location.

On the national security front, combinations A-C and A-D both raise security of supply issues, while combination A-D in particular risks transferring defense-related products and technologies to undesirable parties. If the products and technologies in question are military critical, current U.S. laws and regulations would prohibit both combinations.

Scenario 3: Foreign Ownership/U.S. Research and Development and Manufacturing Location. Foreign investors establish an integrated high-tech subsidiary in the United States; combinations C-E and C-F prevail. In this scenario, American R&D and manufacturing workers share in any excess returns to labor. While excess returns to capital accrue to foreign rather than American owners, it is usually assumed that these excess returns result from knowledge and skills specific to the foreigners and are therefore not "lost" to Americans.[34]

Because both R&D and manufacturing are located in the United States, technological spillovers stay in the United States. Security of supply is also ensured. However, if the foreign investment is in a defense-related industry, steps may be taken to keep the foreign owners from exercising control or influence over decisions involving military-critical items and technologies.

Scenario 4: Foreign Ownership/Foreign R&D and Manufacturing Location. In this scenario, the industry is well and truly lost, with no American ownership and no phase of production under American control. Under combinations D-G and D-F, any rents and any direct spillovers accrue to foreigners. Further, few tools are available to ensure security of supply or to prevent unfriendly powers from obtaining this industry's products or technologies. This is truly the "nightmare" scenario posited by Cohen and Zysman (1987), Joel Popkin and Company (2003), and other critics of hands-off U.S. trade and industrial policies.

However, even under the nightmare scenario, there is a good chance that American consumers would still be better off economically if the benefits of cheaper imports outweigh the losses associated with, for example, R&D and production externalities. In fact, it has been argued that the productivity boom of the mid- to late 1990s—and associated increases in the U.S. standard of living—was largely due to the developments in the U.S. IT industry,

---

[32] Generally speaking, however, decisions to locate overseas reflect a more complex set of considerations. One important factor tends to be increased competition from foreign producers, which erodes profits and therefore excess returns to both labor and capital. As mentioned above, local content requirements—that is, requirements that all or some part of a product be manufactured or assembled in the country where it is to be sold—also affect managers' decisions about where to locate production. See, for example, NRC (1992).

[33] On the other hand, overseas investments designed solely to achieve market access are highly inefficient in terms of global resource allocation because American workers have a comparative advantage in production. From a U.S. policy point of view, the correct action would be to work to eliminate the foreign trade barrier.

[34] Of course, the American owners of firms that compete with foreign-owned but U.S.-located firms may be forced to accept lower returns as a result of increased competition, but their losses would be offset by gains to U.S. consumers.

including the globalization of IT production networks. For example, according to Jorgenson et al. (2003), rapidly falling prices for computer hardware, software, and telecommunications were a major factor contributing to the acceleration in labor productivity growth after 1995.

Conclusion

There is no empirical case to be made that the United States is in danger of losing its overall manufacturing capabilities due to foreign competition. Declines in U.S. manufacturing employment can be directly attributed to increased manufacturing productivity, which has continued to grow since the late 1970s when large U.S. trade deficits in manufactured goods first began to appear. However, in one sense, it appears that Bell was right: As found by workers in developed countries around the world, more Americans than ever now find their employment in the service sector, and their numbers are continuing to grow. But it is not the case that American manufacturing per se is declining. To the contrary, American manufactured goods are more plentiful, and cheaper, than they have ever been before.

In this chapter, we have not presented any empirical evidence that either supports or contradicts the notion that American high-tech industries should receive particular government support or protection. We note, however, that even when theoretically there may be economic justifications for supporting certain industries at the expense of others, the development and implementation of sound industrial policies may be difficult if not impossible. This is because, as argued by Krugman (1996a, p. 111), "you have to base interventionist proposals on detailed predictions about how firms will change their strategies in response to hypothetical policy changes, how these strategic moves will affect profits, wages, R&D, and so on, and finally, how all of these changes will affect the economy at large."

Here, we suggest a simple framework for beginning to think about the different organizational forms that globalization of high-tech industries can take and how they fit within the national security and economic rationales for industrial targeting. For example, several of the studies expressing concern about the health of U.S. high-tech industries argue that, when an industry's manufacturing operations migrate overseas, domestic R&D within that industry suffers (Cohen and Zysman, 1987; OTA, 1988, 1990). One reason offered is that the profitable commercialization of technology—which requires the manufacture of a physical product—is what finances R&D. Another reason is that applied R&D, particularly with respect to process technologies, requires close communication with and proximity to manufacturing to be most effective. If financing is the primary issue, then retaining the U.S. ownership rather than U.S. location of manufacturing is what matters. However, the issue of proximity, which is likely to be industry specific, requires more study.

Another factor to consider when determining whether protection or promotion is needed is the nature of the foreign competitor. For example, in recent years, considerable attention has been paid to India and China as potential high-tech rivals. This represents quite a change from the 1980s and 1990s, when our primary high-tech rival was Japan and firms from India and China were seen as quintessential low-tech competitors responsible for pushing down the wages of unskilled American workers. It is certainly possible that China, for example, is catching up to the United States in state-of-the-art semiconductor manufacturing technology (GAO, 2002). China's very success, however, may demonstrate that certain types of semiconductors have become commodity items and therefore have lost the IRS and beneficial spillover characteristics that justify policy intervention on economic welfare grounds.

In fact, it seems that policymakers' concerns over advancing Chinese semiconductor manufacturing capabilities focus on the implications for U.S. national security rather than for broad-based economic welfare (see, for example, GAO [2002] and Lieberman [2003]). But the case does highlight an area that is quite problematic for policy: dual-use technologies. There is frequently tension between national security–oriented approaches to globalization, in which the preference is for American-owned firms employing American workers to produce items for sale on American markets, and economic approaches, which stress the benefits of trade based on comparative advantage. This tension is unavoidable, and individual cases must be evaluated on their own merits.

In conclusion, it is important to note that industries evolve both structurally and technologically over time. Industries and product lines that once met the definition of "high tech" may not stay that way. For example, DRAM (dynamic random access memory) chips once involved extremely advanced technological processes and highly skilled, specialized labor. In the 1980s and early 1990s, the U.S. government failed to protect its industry against powerhouse Japanese firms. As a result, U.S. firms were forced to move on to products in which they were more competitive, such as microprocessors (Macher, Mowery, and Hodges, 1999). Now DRAM technology is well known, and DRAM chips have become commodity products manufactured competitively and at low cost across Southeast Asia. The many Japanese firms that continued to produce DRAMs are no longer earning rents; in fact, they are now operating at a large competitive disadvantage. The lesson for the U.S. government may be that the "loss" of certain industries to foreign competitors does not necessarily lead to adverse economic or national security outcomes. Further, in the long run, such losses may not be avoidable.

Historical and Structural Effects on Market Share of Semiconductors in the United States

Background

The United States has faced challenges in high-technology manufacturing before. This chapter provides information on one such important case that occurred in the 1980s and describes the policy responses of the U.S. government and their effectiveness. Thus, it provides a historical context for the current debate regarding government support for the U.S. semiconductor industry.

Between 1979 and 1986, the U.S. percentage of the world market for DRAM (the most common type of semiconductor memory) products fell from over 70 percent to under 20 percent and has never recovered. Japanese semiconductor manufacturers had mounted a sustained campaign to capture the market, which they successfully controlled until the emergence of South Korean competition in the 1990s. This chapter describes the structural characteristics of the Japanese and American semiconductor industries and how they contributed to the rise of the Japanese industry in the production of semiconductor memory components. It discusses some of the reasons why the U.S. semiconductor industry, whose worldwide importance was perceived by many as a fading second to the Japanese semiconductor industry in the mid-1980s (Prestowitz, 1988), had regained global preeminence by the early 1990s.

Competition among Japanese and American semiconductor firms is well documented. As early as 1982, in the midst of the ascendance of Japanese firms, Borrus and colleagues published a report detailing the competitive aspects of the industry and the actions taken by the Japanese and American governments to support their respective industries (Borrus, Millstein, and Zysman, 1982). OTA issued a report (*International Competitiveness in Electronics*, 1983) assessing U.S. competitiveness in electronics in general. Prior to that, the patterns of foreign direct investment, which began the worldwide expansion of the industry, were reviewed in several reports by the National Bureau of Economic Analysis (Finan, 1975; Lake, 1976). The global nature of semiconductor manufacturing was discussed in detail in Grunwald and Flamm (1985). Prestowitz (1988) issued last rites to the American industry in 1988, as U.S. semiconductor firms and the U.S. government entered into the SEMATECH (semiconductor manufacturing technology) partnership. Later analyses investigated the causes and characteristics of the decline in U.S. semiconductor memory production in the 1980s and its renaissance in the 1990s, typically assessing it in terms of trade disputes among nations (Tyson, 1992; Krugman, 1992; Wessner, 2003). Still other studies explore the microeconomics of the semiconductor industry, including characterization of its learning curve (Isaac, 2003). Recently, there has been much concern regarding China's entry into semiconductor design and manufacturing and its implications for U.S. national security and industrial competitiveness (GAO, 2002; Lieberman, 2003; Wessner, 2003).

This chapter serves as a literature review of the analyses mentioned above. It discusses the semiconductor industry from its inception in 1958 through the mid-1980s. We recount the emergence of Japan as a significant contributor to semiconductor supply because it is a similar rise of Taiwan or China that forms the basis of current U.S. fears. Table 3.1 presents five aspects of the semiconductor industry on which we will base our discussion of the decline of the U.S. capacity for the production of semiconductor memory products. This table represents the state of the American and Japanese industries from their inception to the mid-1980s, although many of the insights that we glean from the table are still valid today. It might seem simplistic to hang an argument regarding U.S. industrial competitiveness on five characteristics of the respective industries. However, each characteristic can be interpreted in the context of global market share and technological development. For example, U.S. firms, having based their strategic growth on technological innovation, lost significant market share in DRAM production when Japanese 64K chips were produced with failure rates much lower than American chips (Borrus, Millstein, and Zysman, 1982; Prestowitz, 1988). The constant search for new markets and applications for semiconductor devices, which is a characteristic of the U.S. industry, leads naturally to the emergence of semiconductor foundries that specialize in rapidly producing custom integrated circuits (ICs), usually designed elsewhere and for specific applications, in standard manufacturing processes. The presence of a foundry is not prima facie evidence of weakness in the U.S. semiconductor industry but rather an indication that demand across the industry allows certain manufacturing processes to be developed and used at lower collective cost than each individual company could attain; in fact, foundries began to appear in the 1970s (Finan, 1975). This discussion focuses on the structural characteristics listed in the table and interprets historical industrial and market changes based on each of the characteristics.

Table 3.1
Five Structural Characteristics of the U.S. and Japanese Semiconductor Industries Applicable to Analysis from the Early 1960s Through the Mid-1980s

Structural Characteristic	United States	Japan
Industrial character	Dominated by merchant producers	Dominated by large conglomerates
Typical mode of financing	Capital markets	Debt financing
Strategic basis	Technological innovation	Production
Marketing and growth strategies	Constant search for new market opportunities	Domestic interfirm trade
Philosophy regarding role of government	Enforce free trade	Guarantee national industrial competitiveness

The loss of semiconductor memory market share to Japanese competition in the period beginning in the late 1970s and ending in the mid-1980s is our focus. Figure 3.1 illustrates worldwide shipments of DRAM memory devices from 1978 to 1991. Prestowitz (1988) used the decline to make an argument regarding the fall of U.S. industrial competitiveness. He argued that Japanese preeminence in memory production, as demonstrated by the announcement of the 256K chips, and subsequent dominance of the industry throughout the 1980s foretold the decline of the U.S. industry. The U.S. semiconductor market share in all devices fell to 38 percent of the world market in 1988 but rebounded through the 1990s and hovered at approximately 50 percent through 2002 (Hatano, 2003). The U.S. semiconductor industry never regained market share in memory devices; its resurgence was in higher value-added components, such as microprocessors, application-specific integrated circuits (ASICs), and hybrid devices.

Merchant semiconductor firms producing chips for external consumption dominate the U.S. industry. The large conglomerates and electronics systems producers that dominate the Japanese industry have diverse markets to serve, and this characteristic contributed to the slow advancement of the industry (Borrus, Millstein, and Zysman, 1982). American firms relied primarily on capital financing to support their growth, whereas Japanese firms were dependent on debt financing. As a result, Japanese firms were better able to finance growth as technology developed, even through periods of global or local recession. American firms, relying on capital markets, were unable to secure adequate financing for investment in new equipment and R&D during financial downturns. The Japanese applied the same statistical quality assurance methodologies to the production of semiconductors that had fostered their

Figure 3.1
The Share of the DRAM Market for U.S., Japanese, European, and Korean/Taiwanese Firms, 1978–1991

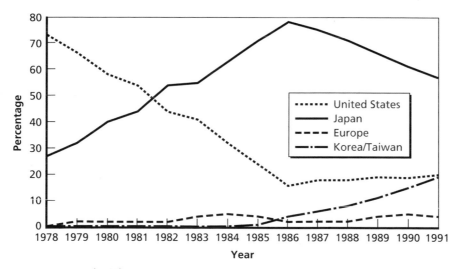

SOURCE: Tyson (1992).
RAND*TR136-3.1*

success in the steel and automobile industries (Prestowitz, 1988).[1] In the case of commodity products, such as memory chips, their focus on quality and yield enabled the Japanese to capture the American market. The American firms organized their strategy around technological innovation, seeking constantly to develop new IC products. The structure of the U.S. industry also led it to seek new markets and products for ICs, hopefully guaranteeing a consistent market for components. The vertically integrated Japanese firms already had a diverse catalog of products of which semiconductor sales composed a small percentage of total revenues. Finally, Japanese government policy, as exercised by the Ministry of International Trade and Industry (MITI), was to guarantee Japanese industrial leadership in strategic industries (Johnson, 1982). The structure of the U.S. industry, while an artifact of the Minuteman missile acquisition and the Apollo space program (Borrus, Millstein, and Zysman, 1982), had developed absent government influence, and its philosophy regarding government involvement was adherence to the principles of free trade.

U.S. Merchant Producers vs. Japanese Conglomerates

Borrus, Millstein, and Zysman (1982) describe the three-tiered structure of U.S. industry. At the top were AT&T's Bell Laboratories (now Lucent) and IBM. These two companies were the largest electronics systems integrators and developers of critical and fundamental technologies for ICs. Below AT&T and IBM were the set of merchant semiconductor firms that provided the industry with its drive to innovate. Since these firms produced chips for external consumption, they actively sought new applications for ICs. Merchant semiconductor manufacturers included (and still include) Intel, Texas Instruments (TI), AMD (Advanced Micro Devices), and others. The third group of firms in the American semiconductor industry comprised electronic systems manufacturers. During the 1970s and early 1980s, these firms included Digital Equipment Corporation (DEC) and Hewlett Packard. Motorola straddled the line between merchant semiconductor manufacturer and systems producer.

The structure of the U.S. industry was an artifact of U.S. antitrust law and choice. AT&T was the inventor of the transistor and held many key patents underlying IC technology. IBM was the largest electronic systems producer. AT&T and IBM did produce ICs, but only for their own consumption. IBM, in particular, incorporated the advancements of the merchant sector into its systems and computing products. The technological leadership of AT&T and IBM combined with their absence from the merchant marketplace had significant consequences:

> Since neither firm competes in the merchant component market—a result of court decree or corporate choice informed by antitrust considerations—both have incentives to trade their technologies for technological developments made by other firms. Such exchanges serve to protect AT&T and IBM against radical breakthroughs elsewhere in the industry, and serve simultaneously to spread their own technological advances to the merchant sector. (Borrus, Millstein, and Zysman, 1982)

---

[1] We note that Japanese quality control methods were adapted from "scientific management" techniques developed by American engineers: The Japanese award the "Deming Prize" for quality control, which is named after American engineer W. Edwards Deming (Johnson, 1982).

The technological leadership of AT&T and IBM, and the technology pull provided by the space industry toward miniaturization, led to explosive and dynamic competition in the merchant sector. Beginning in 1966 and continuing through the mid-1970s, the merchant sector flourished and was a great example of "venture-capital backed entrepreneurship, and the triumph of the technological innovativeness of the small firm" (Borrus, Millstein, and Zysman, 1982). The development of the merchant sector coincided and supported the development of the U.S. computer industry. Table 3.2 below shows the relative growth of nonmilitary IC applications and the rapid growth of the semiconductor industry in general.

The merchant firms were established to serve the growing commercial market for semiconductors. From 1966 to 1976, 36 semiconductor companies were founded in the United States. These firms included National Semiconductor (1966), Intel (1968), Mostek (1969), and Advanced Micro Devices (1969). Intel's contributions during this period are well documented: the introduction of the first general-purpose memory chip (1970) and the microprocessor (circa 1971). Intel's advances were responsible for a period of rapid growth in this industry and the spawning of high-level computing and software programming, which is a component of IT production that we do not consider in this background discussion.

Because this was a period of rapid technological advance, it was also a time of great risk. One example in particular is noted. Much current leading-edge semiconductor technology is based on the complementary metal-oxide semiconductor (CMOS) logic process. But, in the 1960s, several different logic families were in use: Fairchild's ICs were based on resistor capacitor transistor logic (RCTL), and TI used direct-coupled transistor logic (DCTL). In 1965, Fairchild introduced a new series of ICs based on diode transistor logic (DTL) that quickly led the industry. Therefore, from 1964 to 1967, TI witnessed its 32 percent market share in ICs shrink to 16 percent. TI regained its strength with transistor-transistor logic (TTL) in the 1970s. Wild technological change also corresponded to wild swings in market share. The Japanese introduction of high-quality, low-cost memory in the 1980s allowed them to similarly take market share from the American industry (Borrus, Millstein, and Zysman, 1982).

Table 3.2
The Semiconductor Market Grew from Two $4 Million Government Contracts in 1962 to a $2 Billion Market Serving Primarily Computer and Industrial End Uses in 1978

U.S. Markets for ICs by End Use	Percentage of Value in				
	1962	1965	1969	1974	1978
Government	100	55	36	20	10
Computer	0	35	44	36	37.5
Industrial	0	9	16	30	37.5
Consumer	0	1	4	15	15
Total U.S. domestic shipments ($M)	4	79	413	1,204	2,080

SOURCE: Borrus, Millstein, and Zysman (1982).

Global Production

The American semiconductor industry has had an international character since its inception. Semiconductor manufacturing has been a global enterprise since the invention of the IC in 1958. In 1959, TI became the first American semiconductor firm to transfer technology abroad, in this case to Britain; by 1972, American firms had established 108 distinct operations throughout the world (Finan, 1975). Finan's study was significant because it detailed not only the foreign direct investment of U.S. firms but also their desire to expand their market: Thirty-one of the 108 overseas facilities were point-of-sale assembly operations in which the final product was not reimported to the United States (Finan, 1975). In fact, "by 1971 virtually every U.S. semiconductor firm had at least one offshore assembly affiliate . . . . [B]eginning in 1971, the [U.S. Department of] Commerce statistics indicate that the U.S. trade deficit [in ICs] grew steadily from $3 million to $672 million in 1977" (Borrus, Millstein, and Zysman, 1982). However, the semiconductor industry consists of much more than the finished ICs; it also includes wafer processing and assembly and test equipment, as well as the diced chips that become the finished product. When these factors are taken into account, the U.S. industry ran a trade surplus throughout the 1970s. What is germane for our discussion is that the U.S. semiconductor firms have assembled relatively few of their products in the United States since the mid-1970s.

Whereas the American industry was dominated by a large number of merchant semiconductor firms, with characteristically intense competition and rapid technological advancement, the Japanese industry, until the early 1970s, was one in a general stasis. The Japanese semiconductor industry was (and still is) dominated by large multinational electronics firms for which semiconductors comprise a small subset of their product portfolio. Table 3.3 illustrates the major Japanese and U.S. semiconductor producers and the percentage of their 1979 sales attributable to semiconductors. The table presents data for 1979, the eve of Japanese dominance in the 64K DRAM market.

The Japanese firms that came to dominate the semiconductor industry also dominated its consumer electronics industry, and the semiconductor devices that these firms produced were targeted toward these end uses. Integrated circuits were incorporated into consumer electronics in the 1960s, as the Japanese industry consolidated its control over of the

Table 3.3
Semiconductors Are Typically a Small Share of Total Sales for Japanese Producers

Japanese Firms	Percentage of Sales Attributable to Semiconductors in 1979	American Firms	Percentage of Sales Attributable to Semiconductors in 1979
NEC	17.8	AMD	89
Fujitsu	6.7	Fairchild	69
Toshiba	5.5	Intel	75
Hitachi	4.1	Mostek	93
Mitsubishi	3.8	Motorola	31
Matsushita	2.3	National Semiconductor	85
		TI	36

SOURCE: Borrus, Millstein, and Zysman (1982).

world market for consumer electronics. However, the IC demands of the consumer electronics industry were primarily for linear ICs—op-amps, filters, rectifiers, etc.—which required a much lower level of technological development than the logic devices produced by American firms for the domestic U.S. market. In addition, Japanese firms "had a long history of overt and clandestine methods for cartelizing the consumer electronics market in Japan and coordinating export efforts abroad" (Tyson, 1992).

Each major Japanese electronics firm was tied to a *keiretsu*, which is an industrial grouping centered around a major bank that includes a number of large firms and typically a trading company to facilitate sales. The keiretsu structure replaced the family-based *zaibatsu* structure that existed prior to World War II. Throughout the 1950s, MITI's Industrial Rationalization Council helped to structure the keiretsu and assigned the appropriate trading partners (Johnson, 1982). Equity cross-ownership bound keiretsu members to one another (Borrus, Millstein, and Zysman, 1982). While the keiretsu offered its members a steady domestic market for goods, and the trading company facilitated exports, its important characteristic was the access to stable capital for expansion, which we discuss later (Johnson, 1982; Borrus, Millstein, and Zysman, 1982). Table 3.4 lists the major Japanese electronics producers and the keiretsu of which they were members. Hitachi, Matsushita, Mitsubishi, and Toshiba are the dominant firms in their keiretsu.

Broad Technological Innovation in the United States vs. Production in Japan

The Japanese MITI and Nippon Telephone and Telegraph (NTT) in 1975 joined forces and provided approximately $150 million in subsidies to the formation of a joint research program in very large scale integration (VLSI). It is unclear whether the government support took the form of direct payments to the participating firms—NEC, Hitachi, Toshiba, Fujitsu, and Mitsubishi—or loan guarantees. At least $42 million of the project total was spent in the United States to procure advanced semiconductor manufacturing equipment and testing technology (Borrus, Millstein, and Zysman, 1982). The formation of the research group, and the procurement of the most advanced production equipment, signaled that the Japanese were focusing their collective research energy on advanced chip design, not on production technology. At the time, the Japanese semiconductor market for advanced logic devices was small, so the market for these chips was primarily the United States. The MITI- and NTT-financed program promoted a gradual shift in the Japanese semiconductor industry from the production of relatively low-tech linear ICs for its domestic market to the

Table 3.4
A List of Major Japanese Electronics Producers and Their Parent Keiretsu

Firm	Keiretsu
Fujitsu Ltd.	Dai-Ichi Kangyo Bank Group
Hitachi Ltd.	Hitachi Group
Matsushita Electric Industrial	Matsushita Group
Mitsubishi Electric Corporation	Mitsubishi Group
Nippon Electric Company Ltd.	Sumitomo Group
Toshiba Corporation	Toshiba-IHI Group

mature logic devices required by the U.S. computer and high-tech markets (Borrus, Millstein, and Zysman, 1982).

The Japanese efforts were fruitful. In 1975, Mostek released the 16K DRAM, which
was followed shortly by the introduction of the same chip by its U.S. competitors (Prestowitz, 1988). When IBM released the 4300-series microcomputer in 1978, it created a huge
demand for the 16K DRAM, which had just achieved price per bit parity with the previous
generation 4K chips (see Figure 3.2). U.S. producers could not match the demand, and
Japanese producers quickly stepped in. One of the reasons for the lack of U.S. capacity was
that American firms did not adequately invest in production capacity during the 1975–1976
recession, a point on which we elaborate in the next section. The Japanese also brought
another characteristic to the American market: quality. Until the late 1970s, American firms
had primarily focused their efforts on R&D and technological advancement. This was a
rational strategy in a fast-moving market in which a delay of several months in the introduction of a product could eliminate the possibility of profit from the venture: "U.S. manufacturers, until the advent of Japanese competition over quality, had made a tacit decision that
fast, volume output, with component testing to cover imperfections in the manufacturing
process, was more important than high quality" (Borrus, Millstein, and Zysman, 1982). The
Japanese chip manufacturers had focused their efforts on process improvement, yield, and
quality. A few U.S. systems integration firms, "notably Hewlett Packard and NCR [National
Cash Register,] have suggested that the failure rates of the Japanese product were significantly lower than those of U.S. devices" (Borrus, Millstein, and Zysman, 1982). Of course,
higher quality translates to higher costs, but on a product-by-product basis, the Japanese
could not be accused of dumping. By 1979, the Japanese had captured 43 percent of the
U.S. market for 16K DRAMs (Borrus, Millstein, and Zysman, 1982; Prestowitz, 1988).

Figure 3.2
Price vs. Year for Six Generations of DRAM

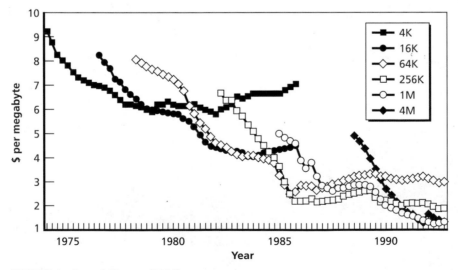

SOURCE: Irwin and Klenow (1994).
RAND*TR136-3.2*

Functionally, DRAMs are among the simpler semiconductor products to design. The Japanese strategy was to capture the market with DRAMs, then use the collective experience gained in producing them to expand into such other semiconductor products as erasable-programmable read-only-memories (EPROMs) and programmable logic devices, eventually dominating the industry technologically (Tyson, 1992).[2] As such, they were following a pattern in which revenues from DRAM sales once allowed the U.S. firms to "test, stabilize, and refine their production and quality-control processes" (Prestowitz, 1988). Throughout the mid-1980s, Prestowitz, the Semiconductor Industry Association (SIA), and the federal government continued to argue that the lessons learned in the production of DRAMs would allow the Japanese a "head start" in the production of different devices. A Japanese scholar of the semiconductor industry, however, observed that "learning economies tend to be product specific because the fabrication process required for one device differs from that for another, and because the photomasks—improvement in which often results in substantial improvement of the yield—are product specific" (Kimura, 1988).

The Japanese were attempting to exploit the "learning curve" for semiconductors. Firms "learn" by lowering production costs as they produce more of a product; the learning rate is the percentage drop in costs for each doubling of output (partially depicted in Figure 3.2). Irwin and Klenow (1994) analyzed the learning process in the production of successive generations of DRAM chips, beginning with the 4K DRAM and ending with the 16M DRAM. They also studied the effect of spillover, which is the amount of learning that occurs due to the advances of another firm. They concluded that

> (a) Learning rates average 20 percent, (b) firms learn three times more from an additional unit of their own cumulative production than from an additional unit of another firm's cumulative production, (c) learning spills over just as much between firms in different countries as between firms within a given country, (d) Japanese firms are indistinguishable from others in learning speeds, and (e) intergenerational spillovers are weak . . . . (Irwin and Klenow, 1994)

The implications of their results were contradictory. Although they demonstrated that firms benefit disproportionately from their own efforts, the sheer volume of the world's DRAM output would tend to mitigate that advantage. In addition, it seemed that there were sufficient technological hurdles between successive generations of chips to allow lessons to translate from one generation to the next. With respect to the Japanese moves to dominate the semiconductor industry via production of DRAMs, the results of Irwin and Klenow show that this strategy was uncertain.

Financing and Insulation from Market Downturns

The Japanese semiconductor industry, like the Japanese steel and automobile industries, benefited from a stable supply of debt capital. The stable supply of capital insulated the Japanese industry from the market cycle and promoted industrial specialization: "The Japa-

---

[2] OTA (1983) disagreed: "The inroads made by Japanese suppliers of commodity-like chips, notably random access memories, portend stronger competition in other types of microelectronic devices but do not translate automatically into advantages for products such as logic chips or microprocessor families."

nese could operate from a strategic point of view, investing heavily even when markets were depressed, so that they would have the capacity to gain a greater share in the next market up-swing" (Prestowitz, 1988). American firms typically financed R&D and growth through capital markets and reinvestment of profits.[3] Whether Japanese capital was actually less expensive than U.S. capital is a matter of debate, but the easy access to investment capital was critical to the growth and technological development of the industry. When coordinated with active government protection of the growing domestic industry from *foreign* competition, the strategy proved remarkably powerful.

In the early 1980s, the 64K DRAM helped the Japanese to further consolidate the gains made in the world market with the 16K DRAM. The Japanese declared victory in the IC industry:

> Any lingering doubts that Japan had caught up with the American semiconductor industry in key areas of advanced semiconductor technology were dispelled early in 1980 when representatives of Matsushita, NEC-Toshiba and Nippon Telephone and Telegraph's Musashino Laboratory provided a solid-state-electronics conference in San Francisco with detailed descriptions of memory chips that could store still another four times as many bits of information—256K—as the 64K RAM chip. At a time when U.S. semiconductor firms were wrestling with 64K RAM production problems, Japanese manufacturers were gearing up for the next generation of 256K dynamic RAMs. (Gregory and Etori, 1981)

Indeed, U.S. manufacturers had designed a 64K chip that included extra rows of memory that could be activated or deactivated during testing; the more complicated design was a technological attempt to increase effective yield (Borrus, Millstein, and Zysman, 1982).[4] The Japanese semiconductor industry, however, produced a much simpler design for the 64K generation, relying on their manufacturing advancements to guarantee yields. In addition to the technological advances, the Japanese dramatically increased their production capacity: "Between July 1981 and August 1982, Japanese capacity for production of 64K RAMs increased from nine million devices per year to sixty-six million" (Prestowitz, 1988). This explosion of production capacity was a direct result of the industrial financial system that Japan used to fund its post–World War II growth. American firms invested heavily between 1981 and 1985—an equivalent of 22 percent of revenues in equipment (Prestowitz, 1988) and 10 percent on R&D (Irwin and Klenow, 1994)—but Japanese producers invested an equivalent of approximately 40 percent of revenues on equipment and 12 percent on R&D over the same period (Prestowitz, 1988).

In 1979, the six largest Japanese semiconductor firms, listed in Table 3.3, controlled 79 percent of the Japanese semiconductor market (Borrus, Millstein, and Zysman, 1982). However,

> Internal consumption by the largest firms of their captive production is relatively low. Approximately 21 percent of the value of production is consumed internally by the ten largest producers . . . . Moreover, internal consumption is particularly

---

[3] For a detailed discussion of the Japanese industrial banking system and practices, see Johnson (1982).

[4] It is not clear whether the more complicated U.S. designs for the 64K DRAM were an attempt to overcome manufacturing shortfalls or an appropriate technological response to the increased density and complexity of the memory chips that followed.

low—10 percent on average among the top four firms in 1979—in MOS devices. Such low internal consumption figures might seem peculiar because these same producers are also the largest *consumers* of domestic semiconductor devices. Indeed, the top ten firms consumed at least 60 percent of total Japanese domestic production, and the percentage of their consumption of the most advanced IC devices is undoubted even higher. (Borrus, Millstein, and Zysman, 1982, emphasis added)

The Japanese firms were at once semiconductor firms and systems producers. Their semiconductor divisions, evidence suggested, had specialized in particular devices or families of devices, while the systems divisions sought needed ICs from other Japanese manufacturers. This de facto product specialization is an example of market rationalization, a process in which competition serves to improve product quality and performance to the benefit of all firms in the marketplace as well as the nation (Johnson, 1982). Over several decades, as firms grew used to trading with one another, the patterns became entrenched and difficult to break as international trade accords were negotiated and introduced (Prestowitz, 1988). Irwin and Klenow (1994) note that "learning by doing is purely internal to the firm, wherein each firm must undertake production itself to reap the cost savings." The concern among some American observers was that spillovers—the effect of one firm's experience on the performance of another firm—was greater in Japan because of the close ties among industry participants.

There was a world recession in 1975 and 1976 that stifled investment in the United States, but because of the debt-based Japanese financing system, it allowed Japan to greatly expand capacity and production. Between 1974 and 1978, Japanese firms increased their production of MOS-based ICs—those for which there was significant demand in the United States—from 46 million to 228 million (Borrus, Millstein, and Zysman, 1982). Support for R&D by MITI and the Japanese government at this time was also significant: Recall the VLSI program, which started in 1975. Additionally, between 1971 and 1980, the Japan Development Bank loaned $1.9 billion to Japanese firms to finance capacity expansion (Borrus, Millstein, and Zysman, 1982). American firms did learn their lesson regarding consistent capacity expansion. Writing in 1982, Borrus, Millstein, and Zysman comment that, since 1975,

> ...the ten largest U.S. merchant producers have more than matched the Japanese in adding new capacity. As a group, these ten producers spent more than $910 million in 1979 and more than $1.2 billion in 1980, or more than 20 percent of sales during each of the two years. (Borrus, Millstein, and Zysman, 1982)

It has already been noted that the Japanese capacity additions in the early part of the 1980s far exceeded those of the American industry.

Technology Transfer and Market Penetration

The investments in capacity expansion, due to the global nature of the semiconductor industry, were not necessarily in the United States or Japan. Early semiconductor manufacturing included an assembly phase that was labor intensive. To exploit cheap labor, chips were shipped abroad, typically to foreign assembly affiliates in Hong Kong, South Korea, Taiwan, Singapore, Malaysia, or Mexico; assembled; and reimported to the United States (Finan, 1975). Two other types of U.S.-owned offshore manufacturing facilities were the point-of-sale (POS) assembly plants that served foreign markets, and the complete manufacturing facility that performed wafer fabrication, assembly, and testing.

Finan (1975) studied patterns of U.S. foreign investment and technology transfer to foreign firms by U.S. semiconductor companies. His study, based on interviews with executives of American semiconductor firms, covered the industry from its inception to the mid-1970s, before the battle over the 16K, 64K, and 256K DRAMs. Figure 3.3 illustrates the expanding pattern of overseas investment by U.S. firms through the early 1970s. Of the 108 facilities, 60 were specifically established to serve the U.S. market, 31 were POS operations, and 17 encompassed complete manufacturing operations serving either the United States or foreign markets (Finan, 1975). Unfortunately, we are unaware of any current studies that illustrate the motives surrounding foreign direct investment by U.S. firms. Foreign assembly became so common that, in the late 1970s, although approximately 80 percent of wafer fabrication occurred in the United States, about 80 percent of assembly and testing occurred overseas (Tyson, 1992). The figure contains all overseas facilities that were in operation in 1974.

U.S. firms had different motives for the establishment of each type of facility. Offshore assembly, whether by affiliate or subcontractor, lowered direct labor costs. Transportation costs for partially complete and assembled semiconductor devices are very small—approximately 5 percent for a mature device—so labor-intensive operations were easily located overseas. Additionally, as yields improved for a particular device, production costs tended to shift toward labor. POS assembly operations were similar to offshore assembly, except they were often located in developed countries. The reason for this is that POS facilities "are located where they can best assist the U.S. firm's penetration of foreign markets . . . two thirds of the POS assembly operations are located in developed countries," and

Figure 3.3
Overseas Operations of U.S. Semiconductor Firms by Year of Establishment

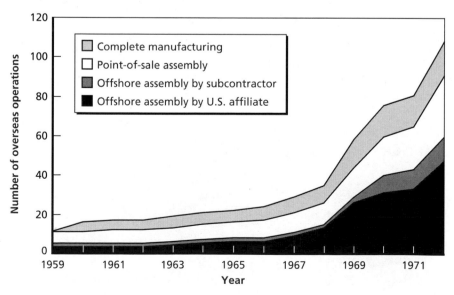

SOURCE: Finan (1975).

RAND*TR136-3.3*

"the ten POS assembly affiliates located in less developed countries primarily export their output to Japan and Europe" (Finan, 1975). Finan's analysis of the motivations behind POS affiliate establishment excluded Japan because "American companies were prevented from having open access to the Japanese market" (Finan, 1975). However, when Finan was writing, Japanese policy allowed reduced tariffs on imports of semiconductor devices from less developed countries; these policies changed in the 1970s (Prestowitz, 1988).

Let us juxtapose our knowledge of the growth of the Japanese semiconductor industry with the pattern of establishment of overseas operations by U.S. firms. Included in Figure 3.3 are four overseas operations in Japan, established in 1968, 1969, and 1972. Consider the 1968 operation as an example. Texas Instruments sought Japanese government approval for the establishment of a subsidiary in the early 1960s. The Japanese government counter-offered with the prospect of minority ownership by TI in a facility. The company refused the offer and used as leverage the fact that NEC had been exploiting TI and Fairchild patents in violation of a licensing agreement. The Japanese government ultimately allowed TI and Sony to jointly own a facility in Japan. TI also agreed to license its IC technology freely to the major Japanese firms: "TI bought Sony's share of the joint venture in 1972, and through 1980 remained the only U.S. merchant firm with a wholly owned manufacturing subsidiary in Japan" (Borrus, Millstein, and Zysman, 1982). IBM had already established a manufacturing facility in Japan but did not compete directly with the merchant firms.

The efforts by other U.S. firms to penetrate the Japanese market were largely unsuccessful, by either POS affiliates or direct imports. Recall that, in 1971, the Japanese began a broad program to develop their domestic semiconductor industry. The strategy of the program was import substitution of semiconductor devices for advanced semiconductor products and continued support for the IC needs of its consumer electronics industry (see the following section). Figure 3.4 illustrates the share of the Japanese semiconductor market controlled by U.S. firms from 1973 to 1986. The TI agreement limited TI to 10 percent of the Japanese IC market, which is approximately the amount of the market that U.S. firms controlled during this period (Borrus, Millstein, and Zysman, 1982; Prestowitz, 1988).

For legal and structural reasons, American semiconductor manufacturers were at a comparative disadvantage in Japan; for Japanese firms, the American market was wide open. Japanese firms established sales and marketing offices throughout the United States during the 1970s and had distribution channels in place in 1978 and 1979 when the U.S. semiconductor manufacturers were unable to meet the demand for 16K DRAM chips; U.S. firms had underinvested in capacity during the 1975 recession (Tyson, 1992). Figure 3.4 shows that U.S. firms increased their share of the Japanese semiconductor market in 1979. So great was the desire of Japanese firms to capture the U.S. market that Japanese firms overexported to the United States and therefore could not meet their own domestic demand, which they then had to satisfy by importing American chips. Of course, since the major producers of semiconductors in Japan were also the major consumers, they were able to recapture their own market and displace the American presence in 1980; American firms, however, held no such leverage over the U.S. market (Borrus, Millstein, and Zysman, 1982). The small increase in Japanese imports ultimately proved detrimental to U.S. firms: Their presence in the Japanese market was ephemeral, while the Japanese established control over the American and worldwide memory market.

Figure 3.4
U.S. Semiconductor Manufacturer Share of the Japanese Semiconductor Market, 1973–1986

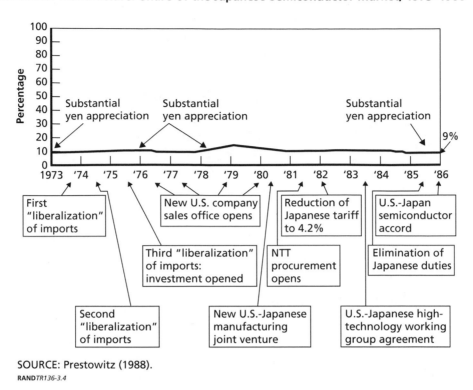

SOURCE: Prestowitz (1988).
RAND*TR136-3.4*

Role of Government, Trade Negotiations, and Market Penetration

The October 1981 issue of *Scientific American* announced the release of the 64K DRAM chip by Mitsubishi and the successful demonstration of the 256K DRAM chip (Gregory and Etori, 1981), which eventually entered the market in late 1984. The Japanese had continued investment in plants and equipment during the early 1980s as the U.S. experienced a recession, which limited the ability of U.S. firms to invest in R&D and equipment (Tyson, 1992). The Japanese introduced 256K DRAM chips before any U.S. firm, including AT&T and IBM. American firms, with the exception of Texas Instruments and Micron (see Table 3.5), ceased to produce DRAMs. The Japanese, eager to capitalize on their market dominance and make up for some of their investment, continued with heavy production. So although prices fell precipitously (see Figure 3.2), the Japanese quickly captured more than 90 percent of the global market for DRAM chips (Prestowitz, 1988).

The collapse of the U.S. DRAM industry occurred under the First Semiconductor Agreement of November 1982 and the Second Semiconductor Agreement of November 1983. The first agreement established a protocol for monitoring shipments of semiconductors but offered no measures to ensure American access to the Japanese market. The second agreement sought to establish permanent and meaningful relationships among Japanese systems firms and American merchant producers (Prestowitz, 1988). The goal of the second agreement was to address the structural problem that existed in Japan in which the major semiconductor producers were also closely linked as consumers of their products. Borrus,

Table 3.5
Producers of DRAM Devices, 1974–1992

Firm	Country	4K	16K	64K	256K	1M	4M	16M
AMD	United States	✓	✓	✓				
AMI	United States	✓						
AT&T-Tech	United States				✓	✓		
Eurotechnique	Europe		✓					
Fairchild	United States	✓	✓	✓				
Fujitsu	Japan	✓	✓	✓	✓	✓	✓	✓
Goldstar	South Korea				✓	✓	✓	
Hitachi	Japan	✓	✓	✓	✓	✓	✓	✓
Hyundai	South Korea				✓	✓	✓	
Inmos	United States				✓	✓		
Intel	United States	✓	✓	✓	✓	✓		
Intersil	United States	✓	✓					
Matsushita	Japan		✓	✓	✓	✓	✓	✓
Micron	United States				✓	✓	✓	
Mitsubishi	Japan		✓	✓	✓	✓	✓	✓
Mostek	United States	✓	✓	✓	✓		✓	
Motorola	United States	✓	✓	✓	✓	✓	✓	
National	United States	✓	✓	✓	✓			
NEC	Japan	✓	✓	✓	✓	✓	✓	✓
NMB	Japan				✓	✓	✓	
Oki	Japan			✓	✓	✓	✓	✓
Samsung	South Korea				✓	✓	✓	✓
Sanyo	Japan				✓	✓	✓	
SGS-Ates	Europe	✓	✓					
Sharp	Japan			✓	✓	✓	✓	
Siemens	Europe		✓	✓	✓	✓	✓	
Signetics	United States	✓	✓					
STC (ITT)	United States	✓	✓	✓				
Texas Instruments	United States	✓	✓	✓	✓	✓	✓	✓
Toshiba	Japan		✓	✓	✓	✓	✓	✓
Vitelic	United States				✓	✓	✓	
Zilog	United States		✓					

SOURCE: Irwin and Klenow (1994).
NOTE: Texas Instruments was the only U.S. firm active in DRAM production for all product generations that Irwin and Klenow studied.

Millstein, and Zysman (1982) observed that these patterns of behavior became entrenched, but writing a decade later, Tyson argued that "the structure of the Japanese industry was obviously more conducive to cartel-like behavior than that of the American industry. Such behavior was also made more likely by the exemption of the computer and semiconductor industries from Japan's Anti-Monopoly Law" (Tyson, 1992).

Japanese industrial policy explicitly had favored the electronics industry since the late 1950s. On June 11, 1957, the Japanese parliament passed, at the behest of MITI, the Electronics Industry Promotion Special Measures Law number 171 (Johnson, 1982; Prestowitz, 1988). In February 1963, the Japanese parliament passed the Special Measures Law for the

Promotion of Designated Industries. Promoted by MITI, it was an evisceration of Japan's antimonopoly law (Johnson, 1982). This law, and several surrounding it, allowed tighter collaboration among Japanese electronics producers as well as favorable access to capital.[5] At that date, the modern Japanese electronics industry had existed for a decade. The earliest technology transfers, from RCA, occurred in the late 1940s and continued through the 1950s. At that time, Japan began to manufacture small transistor radios and electronic parts.[6] Because the transistor was a relatively new technology, the risks were higher and the profits lower than for the established vacuum tube technology produced in the United States. Japanese parts manufacturing helped to bring down costs and make the products profitable. The Japanese used this experience in the 1950s to expand their electronics production into home and automotive audio equipment and, in the 1960s, into semiconductors.

MITI continued its support. Throughout the 1970s and 1980s, MITI invested heavily in joint R&D projects. The VLSI program that helped to build Japanese strength in DRAMs was followed by an $80 million optoelectronics initiative (1979–1986), a $135 million supercomputer program (1981–1989), increased research in VLSI technology (1981–1990 at $140 million), advanced lithography technology (1986–1996 at $62 million), optoelectronic devices (1986–1996 at $42 million), and further research in VLSI technology (1981–1991) (Tyson, 1992).

While MITI explicitly identified the electronics industry for support, and helped to ensure a suitable environment for its growth, the United States did not have a coherent industry policy. In 1981, the Office of Technology Assessment published a report focusing on the steel, electronics, and automobile industries. A critical observation of the report is that the "ad hoc approach to industrial policy followed in years past may not suffice in the current context. Today the United States no longer enjoys the overwhelming technological lead or relative economic strength it possessed two or three decades ago" (OTA, 1981). Regarding goals of industrial policy, the OTA recommended that "macroindustrial policy preserve the flexibility and adaptability of the American economic system while creating a stable climate for industrial growth and the enhancement of competitiveness" (OTA, 1981). The study contained a detailed analysis of the Japanese approach to economic and industrial policy. Writing two years later about the electronics industry in particular, OTA concluded that "it is not realistic to expect that American semiconductor and computer firms will, in the near future and in the absence of cataclysmic changes in other parts of the world, return to the preeminent positions they held at the beginning of the 1970s" and that the U.S. must support its semiconductor and electronics industries "if the United States is to maintain its standard of living, its military security, and if the U.S. economy is to provide well-paying and satisfying jobs for the nation's labor force" (OTA, 1983). The OTA recommended congressional intervention to support the industry.

---

[5] Johnson (1982) notes that Sony and Honda did not establish strong ties (relatively) to the Japanese government yet were very successful in the U.S. market.

[6] These technology transfers were licenses from RCA. Curtis (1994) notes that: "RCA was getting a no-risk royalty return as a percentage of every Japanese radio made and sold in Japan and every set shipped to the United States. Sarnoff [the head of RCA] and his engineers had helped build the new Japanese factories and were continuing to service them with technical aid. The rest of the U.S. radio manufacturing industry, however, was driven out of business." RCA received approximately $100 million annually throughout the 1970s for licensing of consumer electronics technology (Staelin, 1988). License-based technology transfer also occurred in the semiconductor industry in the early stages of the Japanese industry; royalty payments, however, were well documented to industry participants (Finan, 1975).

As the Japanese gained market share in DRAMs with the 256K chip, U.S. firms began to file complaints against the Japanese actions. Despite fierce competition, U.S. semiconductor firms shared many interests and in 1977 formed the SIA to lobby Congress. Complaints regarding Japanese dumping of 64K DRAMs surfaced among SIA members in 1982, but action by the SIA was hampered by the interests of system integrators such as IBM and Hewlett Packard that benefited from a stable supply of inexpensive memory devices. However, the crash in DRAM prices and in the industry spurred by Japanese overproduction of 256K chips instigated a formal complaint by the SIA in June 1985 (Tyson, 1992): The Japanese had been selling the chips for $2 or less when known manufacturing costs were over $3 (Prestowitz, 1988). In September 1985, several firms presented evidence that the Japanese were dumping EPROMs onto the U.S. market (Tyson, 1992). The U.S. Trade Representative publicly contemplated a case against the Japanese throughout the fall of 1985 and decided to proceed in December: "[B]y January 1986 the U.S. government was pursuing a dumping case of its own on 256K RAMs, one dumping petition from private industry on 64K RAMs and one on EPROMs, and a 301 unfair trade case against the Japanese on semiconductors" (Prestowitz, 1988).

U.S. Semiconductor Industry Recovery Post-1986

State of the Industry, 1986–1987

In the mid-1980s, the semiconductor industry was producing a diverse range of products, from simple stand-alone transistors to complicated microprocessors with hundreds of thousands of transistors on a single chip (Yoffie, 1987). These products can be generally classified into three segments: discrete devices such as the single transistor or diode, analog ICs, and digital ICs. In the mid-1980s, the digital market was 70 percent of the total, and memory chips, especially DRAMs, were more than 20 percent of the digital market, or 15 percent of the total semiconductor market (Yoffie, 1987). DRAMs were especially important because they were produced to industrywide standards, making them a true commodity; identical products from any firm could replace those from any other. In contrast, microprocessors, which made up about 10 percent of the digital market in the mid-1980s, were much less interchangeable because they depended on custom software that is exclusive to a particular microprocessor or family of processors. By 1986, Japan had become the largest producer of, and market for, semiconductors, as shown in Figures 3.5 and 3.6. This trend was most apparent in the commoditized DRAM segment, where all but two U.S. manufacturers (TI and Micron) had withdrawn from the market, leaving it to Japanese suppliers such as NEC, Hitachi, Toshiba, Fujitsu, Matsushita, and Mitsubishi.

1986 Agreement

The first U.S.-Japan Semiconductor Agreement, which was formally signed in September 1986, was a response to several actions by the U.S. semiconductor industry and its various members. The history of the agreement has been reported in many publications. Presented here are the key facts as seen by Prestowitz (1988), Flamm (1996), Tyson (1992), and Irwin (1994).

Figure 3.5
Worldwide Semiconductor Revenues by Geographic Market Segment, 1982–2001

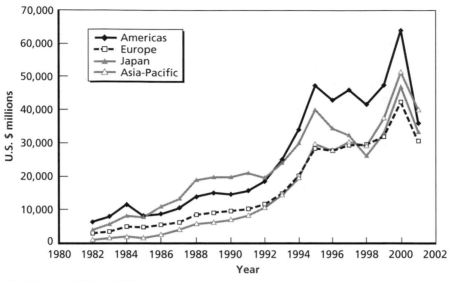

SOURCE: SIA (2002a; 2002b).
RAND*TR136-3.5*

Figure 3.6
Worldwide Semiconductor Market Share by Manufacturer's Home Region, 1982–2001

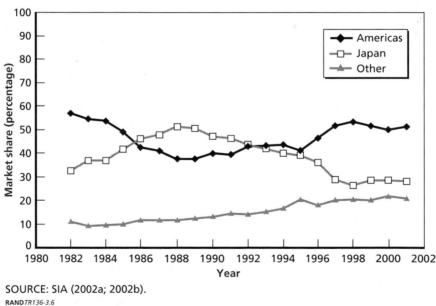

SOURCE: SIA (2002a; 2002b).
RAND*TR136-3.6*

In June 1985, SIA filed a petition with the U.S. Trade Representative (USTR) under Section 301 of the Trade Act of 1974. The petition alleged that there were barriers to U.S. entry into the Japanese market, that the barriers were a structural aspect of the Japanese market, that the Japanese government condoned them, and that Japanese government policy condoned overseas dumping. Shortly after the trade petition, Micron filed an antidumping complaint against Japanese exporters of 64K DRAMS, followed soon thereafter by similar complaints from Intel, AMD, and National Semiconductor—all aimed at Japanese EPROM exporters. This was followed, in December 1985, by the U.S. Commerce Department self-initiating its own 256K DRAM antidumping suit.

In the aftermath of the complaints, and with the knowledge that the Commerce Department almost always finds the existence of dumping, Japan's MITI, on behalf of the Japanese manufacturers, began negotiations with the USTR. While negotiations were under way, from March to May 1986, the Commerce Department reported preliminary findings in the EPROM and 256K DRAM cases, and then a final determination of dumping, together with material injury to the domestic industry, in the 64K DRAM case. In all cases, the findings were for substantial dumping margins, which would inevitably lead to nonnegotiable import duties. Further, and also in May, the U.S. House of Representatives voted 408 to 5 to recommend retaliation against Japan under the Section 301 petition, unless a market access agreement was secured.

Under this mounting pressure, MITI acceded to most of the USTR demands, in a deal designed to settle both the Section 301 and antidumping claims. The agreement had three basic conditions:

- First, Japanese firms would cease dumping in all world markets, which was a substantial precedent in that a bilateral agreement dictated behavior in other markets. Related to this condition, Japanese firms had to develop detailed cost records in order to establish their fair market value (FMV) selling price, which was set at production costs plus 8 percent for profit. The FMV for a firm became a price floor, so the Japanese suppliers could sell at any price at or above the FMV.
- The second condition addressed the market access issue in two ways. In the official agreement, Japan agreed to encourage foreign firms to achieve increased market share in the Japanese market. In a confidential (but now well-known) side letter, the Japanese government "states that it 'understood, welcomed, and would make efforts to assist foreign companies in reaching their goal of a 20 percent market share within five years'" (Tyson, 1992). This would mark a significant gain from 1986, when foreign firms held less than 9 percent of the market.
- In return for these two provisions, the U.S. government agreed to a third condition: Suspend the antidumping duties already levied by the Commerce Department.

Agreement Aftermath

To implement the agreement, MITI imposed an antidumping voluntary export restriction (VER), but it had no statutory authority to force compliance, and there was the prospect for easy evasion by Japanese suppliers. The agreement also led to a quick rise in DRAM prices because the FMV was initially based on old production data. This resulted in a near doubling of the 256K DRAM price and broad dissatisfaction on the part of American electronics manufacturers. Between the continued existence of third-market dumping, and the high prices in the U.S. market, there was strong incentive for arbitrage between the various markets. As a result of these issues, within six months of the conclusion of the agreement, the SIA filed complaints that dumping was still taking place via third-country markets and that there was not a noticeable change in access to the Japanese market. As a result, on April 17, 1987, President Reagan placed a 100 percent tariff on $300 million of Japanese imports of televisions, computers, and power tools, which were all products for which U.S. consumers had alternative sources. Of the $300 million, $135 million was retaliation for injury due to third-country dumping and $165 million for lack of progress in gaining access to the Japanese market. The sanctions persuaded Japanese suppliers to follow the VER, and MITI's actions demonstrated this fact over the next several months. Accordingly, the U.S. government lifted $51 million in sanctions in June and $84 million in November. The $165 million market access tariff remained in place until the 1986 Semiconductor Trade Agreement was renewed in 1991.

In 1987–1988, the U.S. industry was further bolstered by several events in addition to the trade actions. In 1987, the SIA gained government support for the 1988 formation of SEMATECH (a consortium of 14 U.S. chip manufacturers with the goal to support new manufacturing technology development), which was expected to result in superior quality products. The industry members, including Motorola, Intel, Texas Instruments, AT&T, and IBM, and matching government grants from the Advanced Research Products Agency jointly funded SEMATECH. At the same time, rising chip demand, and a rise in the value of the Japanese yen relative to the dollar, provided additional benefit to U.S. semiconductor suppliers. The combination of these events reduced the intensity of the dispute and led the SIA to invite the Electronic Industries Association of Japan (EIAJ) to participate in activities intended to develop long-term relations between U.S. producers and Japanese consumers of electronics. As a result, the EIAJ formed the Users Committee of Foreign Semiconductors and the Distributors Association of Foreign Semiconductors to promote market access opportunities.

In the wake of the accord, the Japanese and remaining U.S. DRAM manufacturers (TI and Micron) made substantial profits due to the raised prices, but there was no reentry of other U.S. manufacturers into the market. There was significant entry of new competition from Korean suppliers such as Samsung and a substantial increase in Japanese market presence of U.S. manufacturers, as measured by expenditures of capital, and personnel. Despite these changes, in 1989, the SIA still maintained that there had not been progress toward the 20 percent market access goal. However, by 1989 the price bubble was costly enough that DRAM customers in the electronics industry, including IBM and Hewlett-Packard, formed their own policy and lobbying group, the Computer Systems Policy Project (CSPP) with views that were contrary to those of the SIA. At the same time, the inauguration of George H.W. Bush as President led the SIA to reappraise the trade situation. At the instruction of the USTR, the SIA and the CSPP entered into negotiations with each other to present a united front in preparation for a renewed agreement at the 1991 expiration of the old one. After a lengthy period, the two organizations announced a joint position that declared the antidumping provisions a success, so the Commerce Department would no longer need to collect FMV data. However, they also agreed that, by the end of 1992, the foreign market share should reach the 20 percent level specified in the 1986 side letter.

The united compromise position of U.S. industry resulted in an easier renegotiation with Japan. In exchange for the United States removing the $165 million in sanctions, the agreement explicitly stated that "The Government of Japan recognizes that the US semiconductor industry expects that the foreign market share will grow to more than 20 percent of the Japanese market by the end of 1992 . . . ." In fact, foreign share of the Japanese market hit 20 percent in the fourth quarter of 1992. The year 1991 also marked the beginning of a sustained five-year industry boom that saw total sales revenue grow to nearly three times the 1990 level. The boom, along with greater R&D and product development associated with it, led the U.S. industry to recapture the worldwide market leadership position in 1993 for the first time since 1985 (shown in Figures 3.5 and 3.6). Since then, U.S. industry has maintained that position, and foreign market share in the Japanese market has approached 30 percent, as shown in Figure 3.7.

Additional Recovery Factors

In addition to formal and informal trade related activities, a number of factors have been cited as causes for U.S. industry's return to leadership in the 1990s. These factors are not independent of each other or of trade activities, but they can be broadly categorized as other government policies, shifts in U.S. manufacturer's product portfolios, improvements in U.S. manufacturing quality, and changes in industry structure. Detailed consideration of issues can be found in Flamm (1996); Macher, Mowery, and Hodges (1999); and elsewhere. A brief summary of this literature is presented here.

Other Government Policies

Until the 1980s, the U.S. semiconductor industry was strongly influenced by concerns deriving from the federal government's antitrust suit against AT&T, which was settled in 1956. One outcome of the suit is that AT&T did not enter the merchant semiconductor market and instead liberally licensed its core technology. As a result, there was little collaboration within the industry. In 1984, the National Cooperative Research Act (NCRA) was passed, and the Justice Department relaxed enforcement of antitrust statutes with respect to determining market power in select industries, including semiconductors. The formation of SEMATECH is, in part, credited to the NCRA.

Figure 3.7
Japanese Semiconductor Market Share by Manufacturer's Home Region, 1982–2001

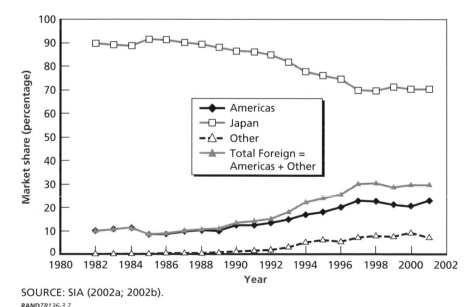

SOURCE: SIA (2002a; 2002b).

The 1980s also marked a shift toward enhanced intellectual property rights. This move is seen in both the 1982 creation of the Court of Appeals for the Federal Circuit, which strengthened patent protection, and the pursuit of international protection during various trade talks. During this change, U.S. manufacturers undertook a dramatic increase in patenting and licensing activities that have led to substantial revenue gains. This swing is generally thought to have aided established firms with strengths in product design and innovation—an area in which U.S. companies have been historically strong in comparison to Japanese firms, which are often thought to excel in quality and manufacturing—but not new product design. The impact of strong property rights on new firms is less clear, given the tradeoff of costs associated with establishing a strong position versus the benefits once it is achieved.

Shifts in Product Portfolio

As already discussed, the recovery of the U.S. industry was not accompanied by any significant growth in the U.S. DRAM market share. Instead, U.S. manufacturers that dominated the memory business of the 1970s and early 1980s, such as Intel, Motorola, TI, and National, have refocused on other products that require more specialized design than memory products. As this occurred, the worldwide revenue in markets for such logic and microcontroller products exceeded the revenue of memory products. In particular, as the Intel microprocessor line (80386, 80486, Pentium, etc.) began to dominate the PC market, the demand for U.S. products grew in Japan and elsewhere. The success of such products has also led to their taking an important role as technology drivers in the development of improved manufacturing processes, equaling or exceeding DRAMs in that function.

While unfair Japanese trade practices, specifically with regard to the manufacture of DRAMs, should have been challenged and were, those practices may have had only a transitory effect on the U.S. semiconductor industry. While the Japanese concentrated their efforts on producing what was essentially a commodity—i.e., DRAMS—the U.S. semiconductor industry focused on where its comparative advantage lies—namely, the design and production of the more technologically advanced and innovative logic and microcontroller products.

Quality Enhancements

Much of the Japanese firms' rise to prominence was attributed to their low price and high quality. In the 1970s and 1980s, the Japanese manufacturers were able to apply quality management practices that they had already developed in other product areas to the manufacture of semiconductor ICs. In 1980, U.S. memory manufacturers had a defect rate that was nearly five times larger than that of the Japanese manufacturers, measured by defective parts per million (ppm) produced. This superior performance enabled Japanese companies to ramp their production lines to high volume more rapidly. However, by the mid-1980s, several leading manufacturers, such as Motorola and TI, had adopted Japanese quality techniques and substantially increased their expenditures on quality assurance. Some industry managers attributed a part of the quality improvement to SEMATECH because the organization enabled better collaboration between suppliers and manufacturers on quality and reliability problems. Between these various efforts, the U.S. manufacturers' defect rate had improved enough to match that of Japanese producers by the early 1990s.

Changes in Industry Structure

The U.S. industry has evolved in many ways since the early 1980s. Of special importance has been the replacement of the stand-alone vertically integrated semiconductor manufacturer with a network of collaborators that specialize in, and excel at, smaller links in the total value chain. This is best exemplified by the rise of the fabless firm, which focuses on design expertise in particular markets and collaborates with foundries that excel at manufacturing high-quality, low-cost products.

The development of this industry model has also led to increased international collaboration with suppliers in third (non-U.S., non-Japanese) markets such as Taiwan that have specialized in foundry manufacturing. This increased collaboration left behind many of the Japanese suppliers that maintained their vertically integrated models, focused on DRAM manufacturing. According to this theory, Japanese manufacturers have been less able to quickly respond to new market demands than has the more distributed industry.

CHAPTER FOUR

Current Computer and Semiconductor Manufacturing Statistics

Trying to understand in quantitative terms what is happening with U.S. manufacturing, and more specifically high-technology manufacturing, is a tall order. As the economic analysis earlier in this report suggests, different perspectives may view the glass as either half full or half empty. To help inform decisionmakers who must weigh policy options, this chapter presents data and statistics on recent high-tech manufacturing trends—specifically computer components and semiconductor hardware manufacturing, the primary focus of PCAST and the subject of much current public discussion.

Beyond high-tech manufacturing, there is a broader concern that U.S. manufacturing in general is disappearing, so data on U.S. manufacturing as a whole are presented to lend perspective of the size of computer and electronics industry within U.S. manufacturing and GDP. Data are thus also presented comparing the employment, value added, and shipment values of the computer and electronics industry, with other major sectors such as transportation, fabricated metals, and food manufacturing.

The rise of the semiconductor industry in Asia has been at the heart of a public debate: What actions must the U.S. government take to stem the migration of an industry that has meant so much to the U.S. economy from moving offshore? To conclude this chapter, data on foreign manufacturing trends are presented—in particular those of China, Taiwan, and other Asian nations compared with U.S. manufacturing statistics.

PCAST also organized numerous meetings with representatives of leading U.S. high-tech companies, state government officials, and others to surface differing experiences and perceptions related to high-tech manufacturing migrating overseas. These meetings were illuminating, for finding out not only which types of manufacturing are migrating overseas but also which types of manufacturing have remained in the United States and thrived. Information from these sessions has been incorporated in the chapter, along with characteristics of U.S. competitive strengths and weaknesses.

Overall U.S. Manufacturing Statistics

To place high-tech manufacturing in context, it is worth examining the overall trends in U.S. manufacturing. Statistics on overall U.S. manufacturing may warrant caution but do not yet portray a crisis. The faster growth of U.S. GDP makes the slower growth of manufacturing an issue in comparison. But in absolute terms, U.S. manufacturing still dominates global production, with value added, wages, and payroll all increasing. Manufacturing employment obviously is the most politically sensitive indicator, and there is certain reason for concern whether foreign competition proves to be the cause. Prior to a dip in 2001 manufacturing employment due to the current economic slowdown, the data show a slow, gradual decline over several decades, which includes the influence of automation and other productivity improvements. The similarly gradual increase in employment in primarily the services sector may also suggest a gradual transition in U.S. industry.

The most detailed data on U.S. manufacturing are from the U.S. Census Bureau's Annual Survey of Manufacturers. The latest report, issued in January 2003, presents data through 2001. As seen in the following figures, the 2001 data show signs of the current recession and economic slowdown. The 2002 and 2003 data will doubtless be of interest when released in the next couple of years. The lag in reporting times, however, is a handicap when analyzing industries that change as rapidly as high-tech industries during a period of economic turmoil. Data from 1993 to 2003 are also presented by the U.S. Department of Labor's Bureau of Labor Statistics (BLS); the data are less detailed than the Census Bureau's but have the advantage of showing more current statistical trends.[1]

The number of U.S. manufacturing jobs has stayed relatively constant for most of the past 50 years with production workers varying between 11 million and 15 million (see Figure 4.1). As discussed in Chapter Two, the *percentage* of manufacturing jobs in the total U.S. workforce has been halved over the same period—i.e., the U.S. workforce has roughly doubled in the past 50 years. Production employees include those engaged in processing, assembling, storing, inspecting, handling, packing, maintenance, repair, janitorial, watchmen service, and working foremen duties. Over the past 20 years, the number of production jobs has been particularly steady, only varying between 11.2 million and 13.5 million.

Figure 4.1
Number of Manufacturing Employees and Production Workers, 1949–2001

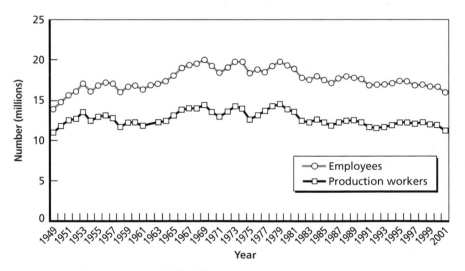

SOURCE: U.S. Census Bureau (1995; 2003).

RAND*TR136-4.1*

---

[1] The BLS manufacturing data for 2003 are an 11-month average from January through November, using preliminary values for October and November 2003 employment.

Recent years, however, are the focus of current public discussion. After 1998, when there were 12.2 million production workers, there began a gradual decrease to 12.0 million in 1999 and 11.9 million in 2000. A sharp drop in production workers then was reported in 2001 during the current U.S. economic slowdown to 11.2 million production workers, or an 8 percent drop since 1998. The BLS data show this downward trend continuing through 2003 to 14.7 million manufacturing employees. These decreases in manufacturing employment may also reflect implementation of the World Trade Organization's (WTO's) Information Technology Agreement (ITA), comprised of four rounds of annual tariff cuts beginning in 1997 and completed in 2000.[2] This agreement benefits U.S. exports of products on the ITA declaration, which includes semiconductor manufacturing equipment and thus may have furthered global production networks for IT firms.[3]

From 1996 to mid-2000, there was also accelerated growth in total multifactor productivity, or total factor productivity, as discussed in Chapter Two. U.S. firms increased output through capital and IT investments and increased use of computers, robotics, and automation.[4,5] The manufacturing output per person has increased by 50 percent in the past decade, while manufacturing employment has gone down (see Figure 4.2). Thus, the difficult question at hand is determining how much of this decrease in manufacturing jobs is due

Figure 4.2
Number of Manufacturing Production Workers and Output per Person, 1993–2003

U.S. Census Bureau (1995, 2003).
RAND*TR136-4.2*

---

[2] For an introduction to the ITA, see www.wto.org/english/tratop_e/inftec_e/itaintro_e.htm (accessed January 2004).

[3] See "Committee on the Expansion of Trade in Information Technology Products," in USTR (2000), available online at www.ustr.gov/wto/99ustrrpt/ustr99_ita.pdf (accessed January 2004).

[4] See Lawrence and Slaughter (1993).

[5] See Lalonde and Lecavelier (2001), available online at www.bis.org/publ/cgfs19boc3.pdf (accessed January 2004).

to the U.S. recession, ITA implementation, increasing productivity, or loss of jobs due to foreign competition.

Manufacturing employees are defined as production employees and those in related support and management functions—e.g., management, personnel, secretarial, sales, and finance. Overall, the number of manufacturing employees parallels the number of production workers and has held relatively steady over the past 50 years. Recent trends for manufacturing employees also follow those of production workers; after 1998, there began a gradual decrease in 1999 and 2000 employees, followed by steep drops through 2003. Since 1995, when there was a recent high of 17.4 million manufacturing employees, there has been a drop of roughly 2.7 million, or 15 percent. During a May 2003 presentation to PCAST, the NAM cited this recent loss of jobs as an area of concern and monitoring.

Data segmenting the U.S. GDP by industry are available from the U.S. Department of Commerce's Bureau of Economic Analysis (BEA). Both manufacturing value added, the dollar value of manufacturing sales minus the dollar value of inputs purchased from other firms, and overall GDP have increased in current dollars over past decades. Manufacturing's percentage of GDP, shown in Figure 4.3, has declined because GDP has grown faster than manufacturing value added. Several sectors have stayed relatively constant since 1987, such as government (both state and federal), retail, transportation, manufacturing, construction, mining, and agriculture, forestry, and fishing. Manufacturing's percentage of GDP has dropped from 18.7 percent in 1987 to 14.1 percent in 2001. In contrast, the services industry, which includes health, business, and legal services, has grown over the same period, increasing from 16.7 percent to 22.1 percent. Other industries that have increased since 1987 are the finance, insurance, and real-estate sectors, which combined have risen from 17.5 percent to 20.6 percent.

Figure 4.3
Percentage of GDP by Industrial Sector, 1987–2001

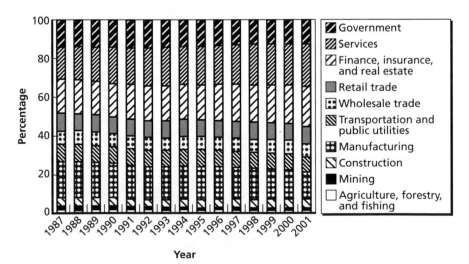

SOURCE: BEA (2002).

RAND*TR136-4.3*

The 25 percent decline in overall manufacturing's percentage of GDP has affected both durable and nondurable goods, which, as shown in Figure 4.4, have declined 26 percent and 23 percent, respectively. Durable goods include lumber and wood products, fabricated metal, industrial machinery and equipment, electronic equipment, and motor vehicles. Nondurable goods include chemical products, food, printing and publishing, paper, rubber, and petroleum products.

Further segmenting the manufacturing percentage of GDP in Figure 4.5, every durable goods product category appears to have shared in the declining percentage of GDP. While overall manufacturing share of GDP has declined by about 25 percent, no particular product category appears to have led this decline. The lumber and wood products industry's percentage of GDP has decreased the most with a 43 percent decline. The primary metals and transportation (other than motor vehicles) industries follow with a 39 percent decrease. The computer industry is included with electronic and other electric equipment whose percentage of U.S. GDP declined by 23 percent. Comparing percentages, however, is difficult and somewhat misleading because the total GDP is growing rapidly. Also mentioned in Chapter Two, the values here are calculated with current-dollar calculations; constant-dollar calculations would be different because the price of products in the construction, transportation, and the service industries rose faster than manufactured goods prices from 1997 to 2001.

Since U.S. GDP has increased faster than manufacturing value added, the percentage contribution of manufacturing to the GDP has decreased. These data will now be presented in constant 1996 dollars, which, as distinguished from current dollars, adjusts for the effects

Figure 4.4
Percentage of GDP for Durable and Nondurable Goods Manufacturing, 1987–2001

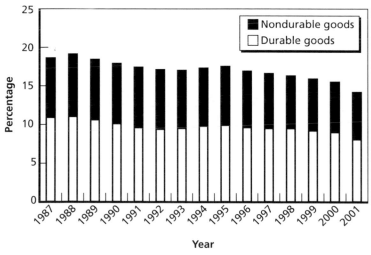

SOURCE: BEA (2002).
RAND*TR136-4.4*

of inflation in the U.S. currency from year to year. As shown in Figure 4.6, the manufacturing, services, finance, insurance, and real-estate sectors combined have actually more than doubled since 1987, with manufacturing output increasing 60 percent over this period. Again, these values are calculated in current dollars, as reported by the BEA, and do not account for the differing rates of price inflation within different industrial sectors and factors needed for calculating constant dollar values.

Figure 4.5
Percentages of Durable Goods Manufacturing Segments of GDP (Computers Included with Electronic Machinery)

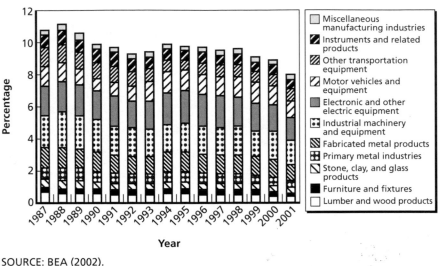

SOURCE: BEA (2002).
RAND*TR136-4.5*

Figure 4.6
Manufacturing, Services, Finance, Insurance, and Real-Estate Contribution to GDP (in 1996 Constant Dollars)

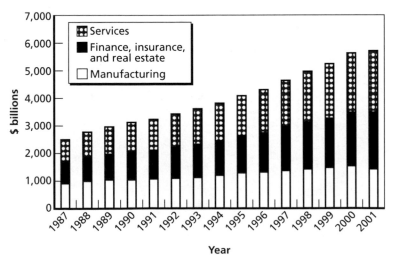

SOURCE: BEA (2002).
RAND*TR136-4.6*

Similarly, Figure 4.7 shows both durable and nondurable goods in inflation-adjusted dollars have increased by more than half since 1987. So while manufacturing's percentage of GDP has decreased, the actual inflation-adjusted contribution of durable and nondurable goods manufacturing has increased.

Figure 4.8 shows a breakdown of durable goods in inflation-adjusted 1996 constant dollars. All product industry segments show modest, if not strong, growth. This includes computer and IT products, which are included in electronic and other electric equipment. The earlier charts displaying declining percentages of GDP are obviously strongly affected by the strong growth of U.S. GDP.

Figure 4.7
Durable and Nondurable Goods Manufacturing Contribution to GDP (in 1996 Constant Dollars)

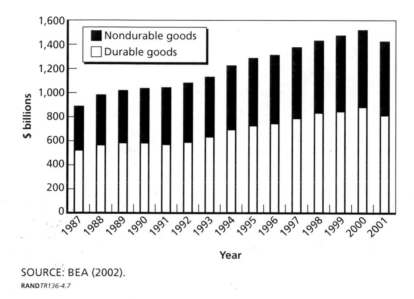

SOURCE: BEA (2002).
RAND*TR136-4.7*

Figure 4.8
Durable Goods Manufacturing Segments Contribution to GDP (in 1996 Constant Dollars)

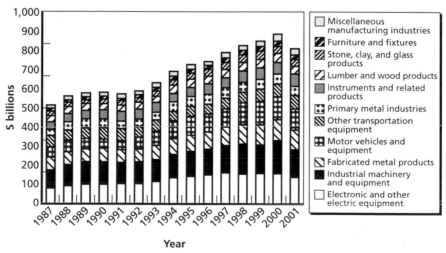

SOURCE: BEA (2002).
RAND*TR136-4.8*

Manufacturing value added, productivity, and contribution to U.S. GDP have grown steadily in current dollars over the past 50 years. However, overall GDP has grown faster, so manufacturing's percentage of GDP has decreased. Table 4.1 shows growth of total GDP, manufacturing, and a breakout of durable goods manufacturing. Motor vehicle manufacturing has led durable goods contribution to GDP with a 91 percent increase since 1987. Lumber and wood products have had the slowest growth, at 22 percent, since 1987. Overall GDP has more than doubled at a 113 percent increase over the same period. As a result, any comparison of manufacturing's contribution (a growth of 60 percent) would be overshadowed by the faster growth of overall GDP.

U.S. Computer and Semiconductor Hardware Manufacturing Statistics

This section focuses on statistical data on current U.S. computer and semiconductor hardware manufacturing. Here, there is more concern than with the overall manufacturing statistics presented earlier, since the high-tech industries are in a much faster period of change. Also, many indicators such as employment and value added per employee showed signs of a decline in 1997–2000 before the economic slowdown in 2001. Causes for this decline include the maturing of high-tech industries and the increasing price sensitivities of its products. Decline in employment in itself is not necessarily indication of a problem, since industrial production in semiconductors, computers, and communications equipment all skyrocketed from 1997 to 2000. However, high-tech manufacturing value added per employee also dropped fast, suggesting its products are rapidly becoming routine or lower-margin, and more attractive targets for foreign manufacturing development.

Table 4.1
Growth and Percentage Change in GDP, Manufacturing, and Durable Goods Segments, 1987–2001

	Growth from 1987 to 2001	Change in Percentage of GDP
Total GDP	113	
Total Manufacturing	60	−25
Durable Goods Manufacturing	57	−26
Motor vehicles and equipment	91	−10
Miscellaneous manufacturing industries	91	−10
Furniture and fixtures	71	−19
Instruments and related products	65	−22
Electronic and other electric equipment	63	−23
Fabricated metal products	61	−24
Stone, clay, and glass products	58	−26
Industrial machinery and equipment	56	−27
Primary metal industries	31	−39
Other transportation equipment	29	−39
Lumber and wood products	22	−43

SOURCE: BEA (2002).

The North American Industry Classification System breaks down overall manufacturing statistics into specific industries, shown in Table 4.2. High-technology industries, and more specifically the IT hardware industry, are concentrated within the three-digit NAICS code 334 for computer and electronic product manufacturing.[6] This includes subcategories for computer and peripheral equipment, which are categorized within 3341, and semiconductor and other electronic component manufacturing within 3344. Given the focus of the PCAST subpanel on computer and semiconductor hardware, statistics for five subcategories will be examined in greater depth, shown in italics and defined below.

Table 4.3 displays those industries categorized by the three-digit level of the NAICS codes and shows computers and electronics products are the third largest employers among U.S. manufacturers. The NAICS code 334 for computer and electronic products includes all the subcategories described earlier (e.g., semiconductor, peripheral, terminal, storage manufacturing). The top five industries account for half the U.S. manufacturing employment. Of the 15.9 million U.S. manufacturing employees, only the fabricated metals and transportation industries employ more people than the computer and electronics industries.

Table 4.2
Relevant NAICS Codes for High-Tech Industries

334 Computer & electronic product manufacturing
- 3341 Computer & peripheral equipment mfg
    - *334111 Electronic computer mfg*
    - *334112 Computer storage device mfg*
    - *334113 Computer terminal mfg*
    - *334119 Other computer peripheral equipment mfg*
- 3342 Communications equipment mfg
- 3343 Audio & video equipment mfg
- 3344 Semiconductor & other electronic component mfg
    - 334411 Electronic tube mfg
    - 334412 Base printed circuit board mfg
    - *334413 Semiconductor & related device mfg*
    - 334414 Electronic capacitor mfg
    - 334415 Electronic resistor mfg
    - 334416 Electronic coil, transformer, & other inductor mfg
    - 334417 Electronic connector mfg
    - 334418 Printed circuit assembly (electronic assembly) mfg
    - 334419 Other electronic component mfg
- 3345 Navigational, measuring, medical, & control instruments mfg
- 3346 Mfg & reproducing magnetic & optical media

SOURCE: NAICS (2002).

---

[6] The SIC system was replaced by the NAICS codes in April 1997. The most obvious difference is the six digits in the NAICS compared with four in the SIC. Both are hierarchal classification systems, with similarities between the higher level industry sectors but more specific segmentation with the additional NAICS digits. Here, higher-level three- and four-digit codes will be used to allow for historical comparisons across the SIC-NAICS transition and more specific codes discussed within the NAICS period.

Table 4.3
Computer and Electronics Manufacturing (NAICS 334) Is a Major U.S. Industry in Terms of Manufacturing Employment, Value Added, and Shipment Value

	Industry Ranking by Size (Percentage of Total)		
	Employment	Value Added	Shipment Value
Fabricated Metals (332)	1st (10.9)	5th (7.5)	6th (6.4)
Transportation (336)	2nd (10.8)	1st (12.2)	1st (15.2)
Computer & Electronics (334)	3rd (10.1)	3rd (12.0)	4th (10.8)
Food Manufacturing (311)	4th (9.5)	4th (10.4)	2nd (11.4)
Machinery (333)	5th (8.3)	6th (7.1)	5th (6.7)

SOURCE: U.S. Census Bureau (2003).

U.S. computer and electronics manufacturing is also the third largest value-added industry, among the three-digit NAICS codes. Total value added for U.S. manufacturing was nearly $2 trillion, with the top four industries accounting for half this total. Transportation goods, which include motor vehicle, motor vehicle parts, and aerospace manufacturing, was the largest, followed by the chemicals and pharmaceutical industry.

In terms of shipment value, the computer and electronics industry (including semiconductors and computer components) is the fourth largest industry in the United States. Transportation, food, and chemicals and pharmaceuticals are the largest three industries. This item covers the received or receivable net selling values of all products shipped, both primary and secondary, as well as all miscellaneous receipts, such as receipts for contract work performed for others, installation and repair, sales of scrap, and sales of products bought and sold without further processing.

From 1997 to 2001, both production workers and all employment positions in the U.S. computer manufacturing industry dropped. The data shown in Figure 4.9 are for NAICS code 3341, which includes the subcategories of electronic computer, computer storage, terminals, and peripherals manufacturing. This mirrors the decreasing trend of overall manufacturing positions. However, the magnitude of the decrease in the computer manufacturing industry is larger than for overall manufacturing. Whereas overall manufacturing employment dropped by 6 percent from 1997 to 2001, computer manufacturing employees decreased by 20 percent over the same period. The decrease of computer manufacturing production workers was even sharper, dropping by 35 percent.

Examining more closely the computer and electronic products industry segments and including semiconductor manufacturing in Figure 4.10, the decrease in employment occurs in all segments. Semiconductor manufacturing has the highest overall employment—about equal to all the computer industries combined.

334111: Electronic Computer Manufacturing. This industry is primarily engaged in manufacturing and/or assembling of electronic computers, including mainframes, PCs, workstations, laptops, and servers. Computers can be analog, digital, or hybrid. Digital computers, the most common of the three, provide the following functions: (1) store the

Figure 4.9
U.S. Computer and Electronics Manufacturing Employment (NAICS Code 3341), 1997–2001

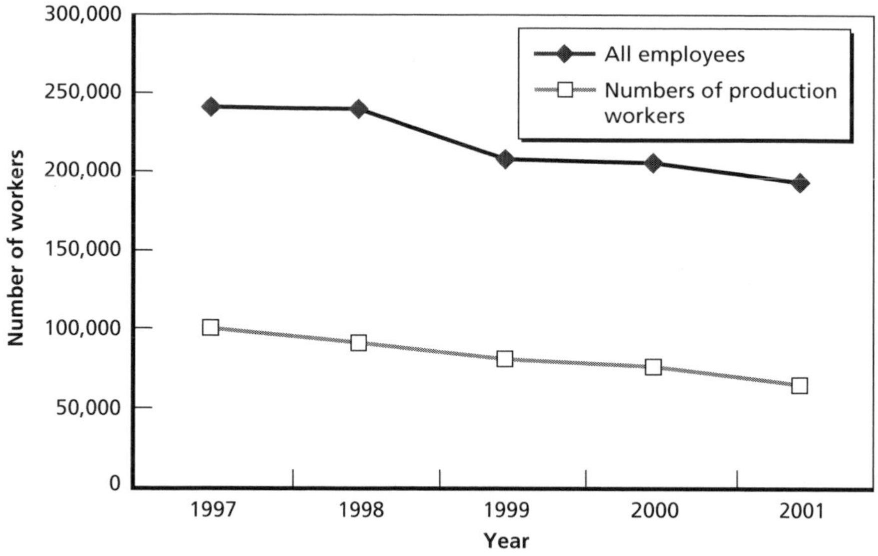

SOURCE: U.S. Census Bureau (2003), data for NAICS code 3341.
RAND*TR136-4.9*

processing program or programs and the data immediately necessary for the execution of the program; (2) can be freely programmed in accordance with the requirements of the user; (3) perform arithmetical computations specified by the user; and (4) execute, without human intervention, a processing program that requires the computer to modify its execution by logical decision during the processing run. Analog computers are capable of simulating mathematical models and contain at least analog, control, and programming elements. The manufacture of computers includes the assembly or integration of processors, coprocessors, memory, storage, and input/output devices into a user-programmable final product.

334112: Computer Storage Device Manufacturing. This industry manufactures computer storage devices that allow the storage and retrieval of data from a phase change, magnetic, optical, or magnetic/optical media. Examples of these products include CD-ROM drives, floppy disk drives, hard disk drives, and tape storage and backup units.

334113: Computer Terminal Manufacturing. This industry manufactures computer terminals, which are input/output devices that connect with a central computer for processing.

334119: Other Computer Peripheral Equipment Manufacturing. This sector manufactures computer peripheral equipment (except storage devices and computer terminals).

334413: Semiconductor and Related Device Manufacturing. This industry manufactures semiconductors and related solid state devices, including integrated circuits, memory chips, microprocessors, diodes, transistors, solar cells, and optoelectronic devices.

Annually, the Federal Reserve Bank releases an industrial production index, which measures the real output of the manufacturing, mining, and electric and gas utilities industries; the reference period for the index is 1997.[7] The latest released statistical data from the Federal Reserve Bank presented in Figure 4.11 show that computers and semiconductor industrial production (relative to levels in 1997) have rebounded since the slowdown in 2001. The logarithmic scale shows over fivefold increase in semiconductor production and a threefold increase in computer production from 1997 to 2003. Communications equipment production, however, peaked in 2000 and has continued to drop through 2003.

The Federal Reserve Bank release also reports on the capacity utilization of industries, the ratio of production to capacity as a percentage of total capacity, shown in Figure 4.12. For a given industry, the capacity utilization rate is equal to an output index (seasonally adjusted) divided by a capacity index. The Federal Reserve Board's capacity indexes attempt

Figure 4.10
Number of Manufacturing Employees in the Computer and Semiconductor Industry, 1997–2001

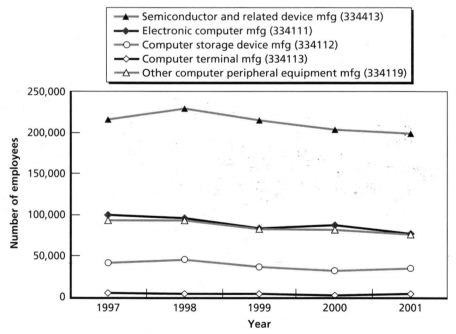

SOURCE: U.S. Census Bureau (2003).
RAND*TR136-4.10*

---

[7] Production indexes for a few industries are derived by dividing estimated nominal output (calculated using unit production or sales and unit values) by a corresponding Fisher price index; the most notable of these fall within the high-tech grouping and include computers, communications equipment, and semiconductors.

Figure 4.11
U.S. Semiconductor and Computer Industrial Production (as a Ratio of 1997 = 100)

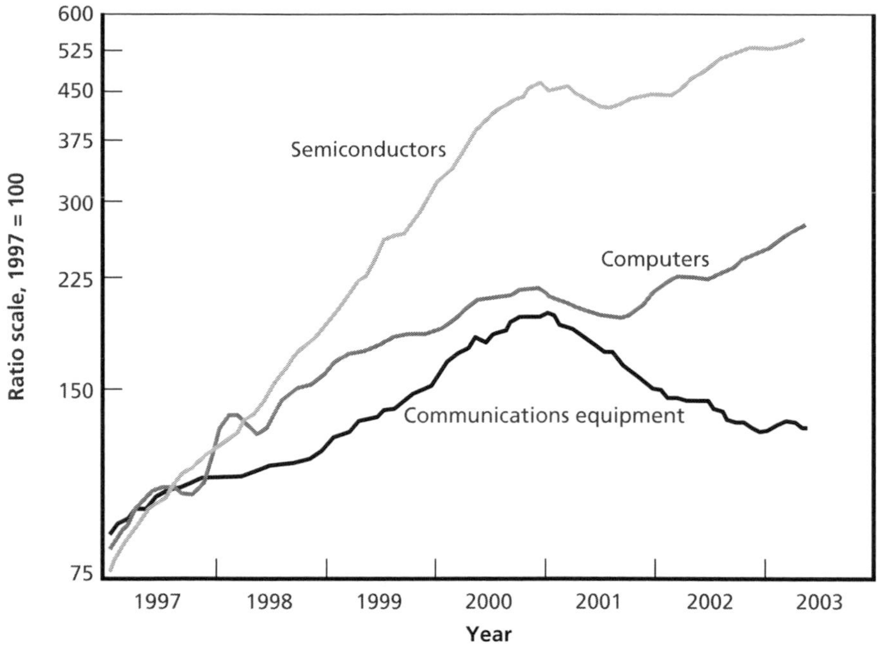

SOURCE: Federal Reserve (2003).
RAND*TR136-4.11*

to capture the concept of sustainable maximum output—the greatest level of output a plant can maintain within the framework of a realistic work schedule, after factoring in normal downtime and assuming sufficient availability of inputs to operate the capital in place. For high-tech industries, capacity utilization dropped sharply after the dot-com boom of 2000.[8] As shown earlier, industrial production and value added of high-tech industries have continued to increase, so this drop may suggest some overcapacity in high-tech industries.

In the last year since May 2002, computer and electronics manufacturing has made the largest percentage change of industries reported by the Federal Reserve Bank, shown in Figure 4.13. Nonmetallic mineral products and computer and electronics manufacturing were the only industries to improve industrial production from May 2002 to May 2003.

As shown in Figure 4.14, which presents data from the U.S. Census Annual Survey of Manufacturers, the computer and electronic product industry (334) overall has similar value per employee as industries such as aerospace, paper, and food manufacturing. This categorization is very broad, however, aggregating many different computer and electronics manufacturing with very different value added per employee.

---

[8] The Federal Reserve Bank defines high-tech industries to include the manufacturers of semiconductors and related electronic components (NAICS 334412-9), computers (NAICS 3341), and communications equipment (NAICS 3342).

Figure 4.12
Capacity Utilization of High-Tech Industries

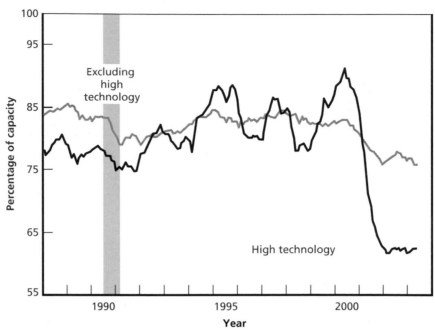

SOURCE: Federal Reserve (2003).
NOTE: The shaded area is a period of business recession as defined by the National
Bureau of Economic Research.
RANDTR136-4.12

Figure 4.13
Manufacturing Industrial Production Percentage Change, May 2002–May 2003

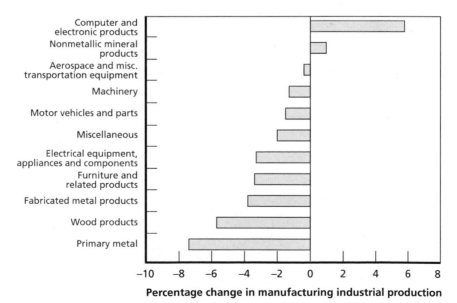

SOURCE: Federal Reserve (2003).
RANDTR136-4.13

Examining more closely computer manufacturing by its subcategories in Figure 4.15, the value added per employee shows two distinct groups. Semiconductor, electronic computer, and storage device manufacturing all have valued added per employee between $200,000 and $227,000. However, hand computer terminal and peripheral manufacturing are significantly lower at $114,000 and $117,000, respectively.

Figure 4.14
Computer Manufacturing Value Added per Employee Compared with Other Major Industries

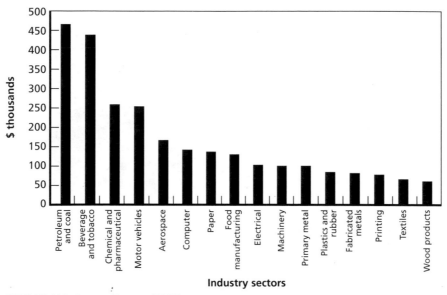

SOURCE: U.S. Census Bureau (2003).
RAND*TR136-4.14*

Figure 4.15
Computer and Semiconductor Manufacturing Value Added per Employee

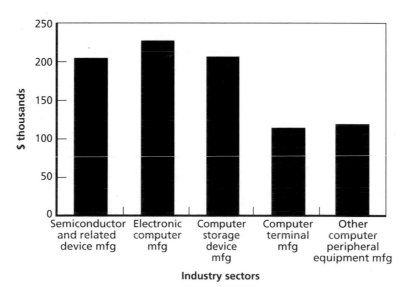

SOURCE: U.S. Census Bureau (2003).
RAND*TR136-4.15*

The value added per employee for the high value added per worker segments has dropped sharply in recent years, when calculated using the numbers of employees and current dollar values in the U.S. Annual Survey of Manufacturers. As shown in Figure 4.16, computer storage device and peripheral equipment manufacturing also increased with the technology boom through 2000, but decreased in 2001. Electronic computer manufacturing increased and peaked in 1999, but began its sharp decrease earlier in 2000. Value added per employee in semiconductor manufacturing also experienced a sharp drop in 2001, following smaller drops in 1998 and 2000. As discussed in Chapter Two, falling prices in the IT sector have resulted in part from the increases in U.S. total multifactor productivity since 1995, as well as globalization of the industry. These issues also factor directly into decreased value added for manufacturing, and correspondingly decreased value added per employee.

Additional evidence of the falling prices of IT products shows up in the decreasing value of shipments of manufactured technology products, shown in Figure 4.17. Correspondingly, decreasing shipment values have contributed to the sharp decrease in value added per worker in U.S. computer and semiconductor manufacturing. Semiconductors and electronic computer manufacturing have been the industries with the highest total value of shipments. However, both experienced a sharp drop in shipment value in 2001, and the electronic computer manufacturing industry began its decline earlier in 2000. The computer

Figure 4.16
Value per Employee for Computer and Semiconductor Manufacturing, 1997–2001

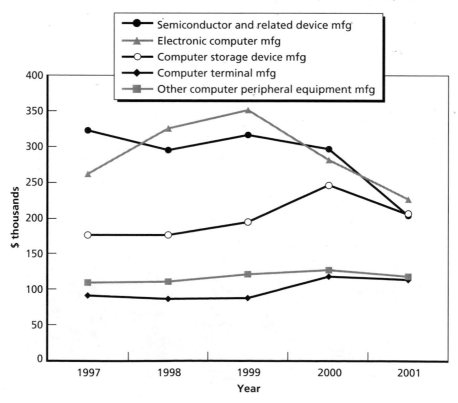

SOURCE: U.S. Census Bureau (2003).
RAND*TR136-4.16*

Figure 4.17
Value of U.S. Computer and Semiconductor Manufacturing Shipments, 1997–2001

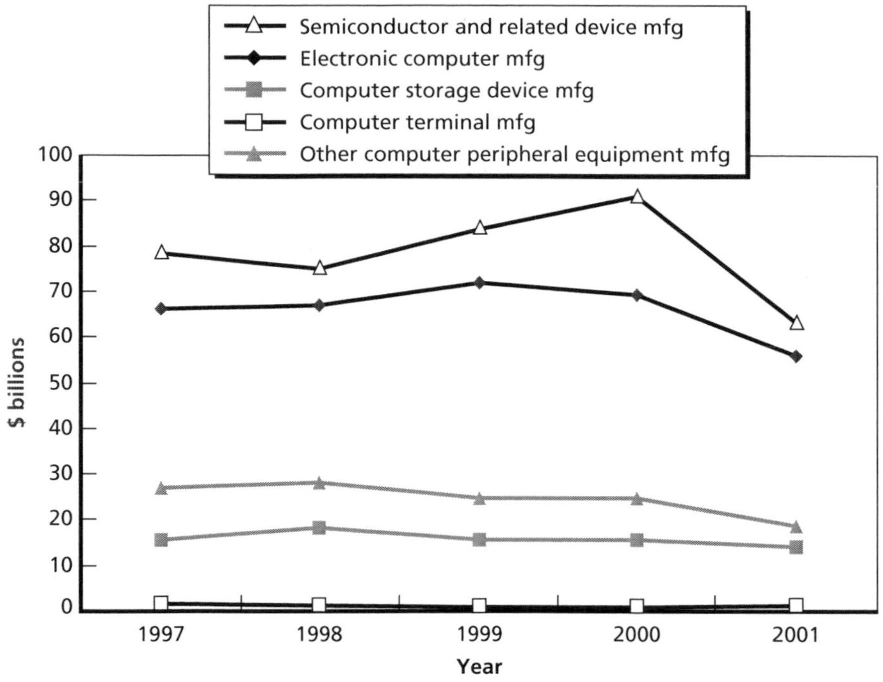

SOURCE: U.S. Census Bureau (2003).

RAND*TR136-4.17*

storage, terminal, and peripheral manufacturing segments are all much smaller in size but have been decreasing as well, beginning their declines earlier in 1998.

Economic Clusters and Regional Economies

Arguments for retaining a U.S. manufacturing base for high-tech products have cited the need to maintain geographic proximity between manufacturing and other supporting activities in the United States. The theory is that manufacturing, along with corporate R&D, university research, educated and skilled workforce, and consumers, are all components of an interconnected "innovation ecosystem." Working properly, the parts of the ecosystem such as great universities and great companies might complement and enhance each other to generate leading manufacturing, innovation, and products for consumers, which in turn spurs new ideas for researchers and entrepreneurs.

The current concern is that the loss of manufacturing from the ecosystem may be detrimental to the effectiveness of federal or industry R&D or that students will be discouraged from pursuing science- or technology-relevant fields if the perception is that high-wage, high-tech jobs have moved overseas.

There are many economic and business articles on the clustering of industries in specific regional areas. These include the clustering of entertainment and aerospace industries in Southern California, the U.S. automotive industry surrounding Detroit, high-tech ventures around San Jose, financial services in New York City, or biotechnology firms in the Route

128 area surrounding Boston. This is not to suggest that these industries are limited to any particular area but that there are concentrations of industries associated with specific regions of the country.

Two analyses on regional economies and industrial clustering are discussed in recent books by Kenichi Ohmae and reports by Michael Porter.[9] The benefits of regional economies and clusters of innovation, such as in the regions listed above, are incentive enough to try to duplicate their success. Would it be possible to create another Silicon Valley or Route 128? Both authors suggest it would be very difficult.

Porter et al. analyzed five regional studies to identify factors such as geography, climate, and population—with entrepreneurship, the presence of research and training institutions, the composition of the regional economy, and public- and private-sector actions as important influences. His research notes that successful regions do not pick winners; they build on inherited assets, and the unique mix of assets may take decades to form a specialized cluster. Porter also remarks that high-tech clusters account for a small percentage of jobs and wages in most regional economies. While communications equipment, analytical instruments, biotechnology/pharmaceuticals, and information technology can be very productive and pay high wages, the overall impact of these clusters on a regional economy is usually relatively small. Porter states that national governments should play a leadership role in developing regional cluster economies, starting from the K–12 education system, support of university and specialized research centers, and setting forth a positive business environment.

Ohmae is more philosophical in his analysis of regional economies, contending that four forces—capital, corporations, consumers, and communication—are the main ingredients for their formation. Beyond examining regions of the United States, he observes other prosperous regions of the world, such as northern Italy, Hong Kong/Southern China, New Zealand, and the growth triangle of Singapore. The free flow of investment and industry, without regard to foreign or domestic origin, with individual consumers empowered by information technologies, are claimed to be the power behind the economy. IT and free communication help consumers decide what to buy, industries decide what to build, and investors decide what to capitalize. For instance, Silicon Valley's historical laid-back, free-wheeling style attracted top-flight people, ideas, and venture capitalists and allowed them to combine and recombine, as business opportunities emerged. For this reason, the national government role is much diminished to Ohmae, since its intervention may hinder the energy of (relatively) unfettered interaction among individuals, industry, investment, and information.

Proximity with Consumers

Some business functions such as marketing, advertising, and sales traditionally benefit from being close to the customer, where nuances in customer patterns and preferences can be closely monitored. There is little to suggest, however, that *manufacturing* of high-tech products needs to be performed near U.S. consumers. High-tech goods generally are not perishable, at least not on the timescales for modern logistics to move them globally. U.S. consumers also have many decades of familiarity with purchasing products made abroad, whether

---

[9] See Ohmae (1995) and numerous reports by Michael Porter at the Council on Competitiveness at www.compete.org/publications/clusters_reports.asp (accessed January 2004).

they be textiles, automobiles, or electronic products. When a domestic and imported foreign product are otherwise comparable, however, two factors may affect which is purchased.

First, the "Buy American" Act of 1933 (41 U.S.C. 10a–10c) was designed to ensure that, when taxpayers' money is spent on direct federal government procurement and infrastructure projects, such expenditures preferentially stimulate U.S. production and job creation.[10] The implementation of Buy American is complicated, however, by debate over what constitutes American products when foreign components are incorporated. Executive Order No. 10582 (1954) defines materials as "of foreign origin" if the cost of the foreign products used in such materials constitutes 50 percent or more of the cost of all products used in the materials. In 1982, an attempt to protect the U.S. auto industry failed to pass legislation requiring 90 percent of components to be U.S. manufactured. The Federal Acquisition Streamlining Act of 1994, while providing exemption of purchases $2,500 or less, changed the "50 percent components test" to a vague standard of "substantial transformation." Debate has continued on the merits of Buy American when foreign acquisition may be more cost-effective or where global markets exist. For instance, the recent Defense Federal Acquisition Regulation Supplement on foreign acquisition (Part 225, revision December 20, 2002) provides an exception to Buy American to ensure access to advanced state-of-the-art commercial technology.

Second, patriotism should not be discounted as a factor to buy a domestic product over an imported product. During periods of U.S. public debate, Congress and others in the past have urged citizens to "Buy American"—e.g., during the 1980s debate over the trade deficit with Japan. There is some dispute over whether "Buy American" is equivalent to "Made in America," the former distinguishing between U.S. and foreign corporate ownership regardless of where products are manufactured. Such public awareness may have played a role with the decisions by Honda and Toyota to establish auto manufacturing in the United States and almost certainly led to advertising campaigns touting their cars as "Made in the USA."

The issue of manufacturing proximity to consumers also applies to U.S. companies that want to establish legitimacy abroad. Setting up a manufacturing plant overseas is viewed by some corporate executives as a way to increase corporate presence and market access, particularly if there is a foreign equivalent of the "Buy American" Act for products manufactured overseas.

Proximity with Universities

There are several studies to suggest that universities can be the catalyst of high-tech innovation, particularly when surrounded by other complementary institutions. Technology transfer of university innovations, many the result of federal R&D funding, has also been an area investigated by PCAST and S&TPI (Wang et al., 2003). Geographic proximity to university research has substantial effect on corporate patent activity in drugs, medical technology, electronics, optics, and nuclear technology, and less in other fields.[11]

---

[10] This summarizes information from Smyth (1999), available online at www.dau.mil/pubs/arq/99arq/smyth.pdf (accessed January 2004).

[11] Link and Rees (1990); Audretsch and Feldman (1996); Jaffe (1986); Jaffe (1989); Cohen and Levinthal (1989); Acs, Audretsch, and Feldman (1994).

A Southern Growth Policies board report, *Innovation.U*, of 12 major research universities noted that successful commercialization or technology transfer of university innovations was enhanced by nearby startup incubators, venture capital investors, and connections with local industry.[12] Universities could enhance these factors by cultivating industry, investor, incubators connections, as well as interfacing with state and regional economic development activities. Of the 12 universities profiled in *Innovation.U*, most come from regions dominated by traditional heavy manufacturing or agricultural industries (Virginia Tech, Georgia Tech, University of Utah).[13] North Carolina State University, Stanford University, and the University of California at San Diego are highlighted as the three institutions from regions that have completed the transition from durable commodity manufacturing to a more technology-driven one.

Stanford University is, of course, at the heart of California's center of entrepreneurship and innovation—Silicon Valley—complete with many startup incubators and venture capital institutions. In an area smaller than Rhode Island, Silicon Valley was estimated to have 7,500 high-tech companies in 1998 (Sherwin, 1998). Metropolitan Boston's eight major research universities are also credited with $7 billion in annual economic impact.[14] This report included Boston College, Boston University, Brandeis University, Harvard University, Massachusetts Institute of Technology, Northeastern University, Tufts University, and the University of Massachusetts at Boston.

Proximity with Capital

Proximity with capital is not actually suggesting that physical money needs to be near manufacturing plants. Rather, the decisionmakers who control the capital, whether they be venture capitalist investors or corporate executives, seem to value proximity with their investments.

Venture capitalists sometimes use the term "smart money" to denote those investors who can coach entrepreneurs and support ventures through their connections and industrial knowledge, as opposed to "dumb money" where little other assistance accompanies the financial investment. Smart money venture capital firms thus prefer to invest in nearby startups, where they can intervene quickly and directly, concentrating investments in a small geographic area.

In its presentation to PCAST, Texas Instruments discussed the construction of its latest facility in Richardson, Texas, and cited the proximity to both TI headquarters and existing infrastructure. Similarly, IBM is establishing a state-of-the-art facility in Fishkill, New York, an hour outside corporate headquarters in White Plains. Both site selections were made with significant incentives from the states of Texas and New York, but the proximity with the corporate headquarters is unlikely a coincidence.

---

[12] See the Southern Growth Policies Board (2002, p. 15), available online at www.southern.org/pubs/innovationU/ (accessed January 2004).

[13] The other institutions profiled include Carnegie Mellon University, North Carolina State University, Ohio State University, Pennsylvania State University, Purdue University, Texas A&M University, and University of Wisconsin.

[14] See AICUM (2003).

Global Information Technology Manufacturing

This section presents data gathered to date on foreign manufacturing of high-tech products, in comparison to U.S. manufacturing. It may be difficult to determine whether foreign advances have come at the expense of, or to the mutual benefit of, the United States. The answer depends on whether the perspective is that of a manufacturing employee, corporate stockholder, government official, or a consumer of high-tech products. U.S. companies are also some of the manufacturing and market share leaders overseas, necessarily seeking lower labor and other costs as a competitive advantage in price sensitive markets. So while the rise of foreign high-tech manufacturing is unmistakable, the implications and policy actions are much less clear.

Figure 4.18 presents high-tech exports in 2000, here defined beyond computer components and semiconductors to include electronics and communications products, for the U.S. and leading foreign producers. These data are from *The Information Revolution in Asia* by Hachigian and Wu (2003).[15] It reported that Asia accounts for more than 80 percent of the total world output of the following IT products: desktop PCs; notebook PCs; cathode-ray tube (CRT) monitors; flat-panel displays; modems; network interface cards; HDDs; computer mouse devices; keyboards; televisions; game boxes; mobile phones; personal digital assistants (PDAs); entry-level servers; hubs; and switches. For the world's semiconductor industry, Asia produces over 70 percent of all bare silicon material, over 90 percent of epoxy resin for IC packaging, over 80 percent of memory semiconductors (DRAM, SRAM [static random access memory], and flash memory), and over 75 percent of outsource manufactured semiconductors.[16]

The manufacture of computer displays may have an especially informative history, if the current growth of high-tech manufacturing overseas is such a concern that the United States would relinquish manufacture of emerging technologies. In the early 1960s, Japan cornered both the market and manufacturing share of televisions, making Sony and Panasonic household names (DeLong, 2001). The first patented aperture grille televisions were manufactured by Sony under the Trinitron brand name, which the company carried over to its leading line of CRT computer monitors. The major competing "shadow mask" technology was actually developed by RCA in 1953 but was later improved and overtaken by Japanese companies. Eventually, shadow mask CRT computer monitor manufacturing became synonymous with products by NEC and Panasonic, many manufactured in Taiwan. The Japanese dominance in television manufacturing, which gave a head start in computer CRT manufacturing, has seemingly also led to dominance in the latest flat-panel display products. In 2002, 79 percent of the plasma display panel systems were manufactured in Japan, 17 percent in the Far East (Korea and Taiwan), and 4 percent in Europe (iSuppli Corp., 2003). The major U.S. manufacturer of flat-panel displays is Corning Display Technologies Corp., a leading supplier of glass substrates for liquid crystal displays (LCDs), has located plants in the major LCD manufacturing centers of the world to be closer to other partnering suppliers and manufacturers; these centers are in Tainan, Taiwan; Shizuoka, Japan; and two Samsung Corning Precision plants in Gumi and Asan, South Korea (Corning Corp., 2002).

---

[15] It might seem odd that Singapore's high-tech exports are more than 100 percent of GDP. GDP is calculated on a value-added basis, but exports are calculated on the estimated shipping value of the finished product.

[16] Data from Taiwan Market Intelligence Center press releases.

Taiwan's electronics contract manufacturing industries face their own challenges, competing with lower labor cost countries or increasing manufacturing productivity (Bout, 2003). According to *Asiaweek*, China has overtaken Taiwan as the world's third-largest manufacturer of information technology products and is gaining fast on no. 2 Japan (Gay, 2001). Semiconductor chip production is expected to increase by 42 percent annually, compared with the world average of 10 percent (Chen and Woetzel, 2002). Like other Asian nations, most of the manufacturing gains so far have been at the low-tech end of the product spectrum. These gains have also been fueled not just by U.S. companies setting up operations in China but also by those headquartered in Taiwan and elsewhere in Asia. Victor Tsan, managing director of the Market Intelligence Center, predicts that within five years, 80 percent of Taiwan's technology output will be made in China (see Chapter Seven on investment flows between Taiwan and China).

China has been the subject of so much discussion regarding accelerated growth of its high-tech industries. However, its GDP still trails that of the United States by a significant margin as shown in Figure 4.19. Projections of how soon China will match or surpass the United States are subject to many factors over a likelihood of several decades.

Discussions on U.S. manufacturing employment have raised the question of whether jobs have migrated overseas—i.e., whether jobs lost in the United States emerge elsewhere. Figure 4.20 shows China's overall manufacturing employment and a subcategory of managerial staff and administrative workers (called "staff and workers" in China's statistical measurements). From 1978 through 1995, total manufacturing employment in China grew rapidly before leveling at just fewer than 100 million employees. Staff and workers also grew from 1978 but reached a plateau earlier around 1992. In 1998, however, China reported a steep drop of 13 million staff and workers during a massive overhaul of the country's state-

Figure 4.18
The Value of U.S. High-Tech Exports Far Surpassed All Foreign Countries in 2000

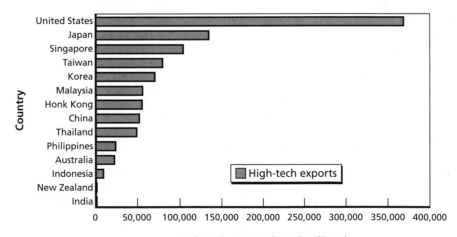

RAND*TR136-4.18*

Figure 4.19
GDP for China and the United States

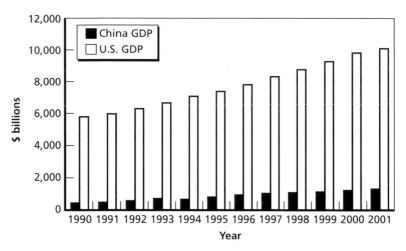

SOURCE: China GDP from Sherman and Jones (2002); U.S. GDP from BEA (2002).
RAND*TR136-4.19*

Figure 4.20
Manufacturing Total Employment, and Staff and Worker Employment, in China, 1978–2002

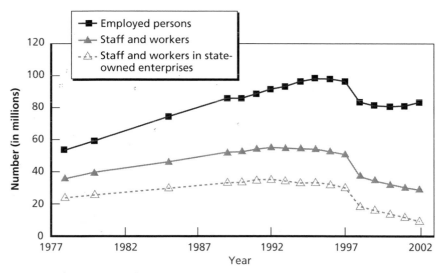

SOURCE: China Statistical Yearbook (2003), data compiled by the National Bureau of
Statistics of China.
RAND*TR136-4.20*

owned enterprises. Numbers of staff and workers have continued to drop through 2002,
again attributed to continued deep reductions in staff and workers at state-owned enterprises.
Since the total manufacturing employed persons leveled off at around 80 million, the
numbers of China's nonmanagerial and nonadministrative workers—i.e., production
workers—have increased roughly by 9 million from 1998 to 2002. Data in 2002 show
roughly a 2 million person increase in manufacturing employed persons over 2001.

However, the data available were not segmented by industry or specific high-tech sectors to allow discussion or comparison with U.S. NAICS segmented manufacturing surveys.

Foreign direct investment (FDI) in China has contributed to its emergence as a manufacturer of high-tech goods. Throughout the 1990s, FDI in China was second only to that of the United States. Foreign investment and joint ventures have been dominated by Chinese-speaking countries around the Asian rim, such as Hong Kong and Taiwan. FDI in China increased significantly in the early 1990s, peaking in 1993 (shown in Figure 4.21). Since then, it has dropped off, leveling at around $55 billion per year since 1997. Much of the FDI listed as being from Hong Kong is actually foreign capital funneled through Chinese intermediaries and partners.

One of the reasons high-tech companies are attracted to Asia is the large size and steep growth of the Asian Internet market (shown in Table 4.4). In 2000, the United States led the world in Internet users. However, Japan was not far behind with about half the Internet users as the United States, but with a year-on-year growth rate more than double that of the United States. China had a quarter the number of Internet users as the United States, but its growth rate was five times more than that of the United States.

Worldwide wage rates for direct labor electronics manufacturing may have a significant influence in corporate decisions on where to locate manufacturing plants. To account for differences between countries, the average direct hourly rate is multiplied by a social costs percentage and overhead burdens to calculate a fully burdened rate in a country. Figure 4.22 shows the different fully burdened rates in 2001 dollars for different countries. The United States at $43.20 per hour is not among the highest for electronics manufacturing wage rates;

Figure 4.21
Foreign Direct Investment in China Peaked in 1993 and Has Been Level Since 1997

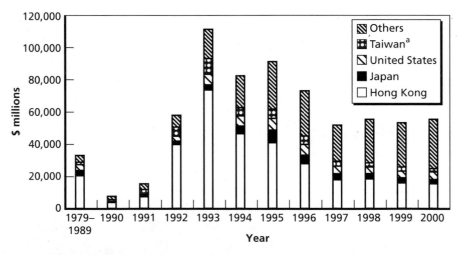

SOURCE: Sherman and Jones (2002).
[a]Prior to 1992, MOFTEC did not track Taiwan investment as FDI. Taiwan data from 1987 to 1991 are estimates from U.S.-China business council files.
RAND*TR136-4.21*

that distinction belongs to Germany and Switzerland, with electronics leader Japan third at $61.54 per hour. The United States is actually near the world average of $42.19 per hour for direct labor wage rates in electronics manufacturing. However, Taiwan, at $16.40, is only one-third of the U.S. wage rate, and China, at $8.50, is one-fifth the U.S. rate. For manufacturing that is labor intensive, these differences can dramatically reduce labor expenses and sway corporate site selection decisions toward lower-cost countries.

Table 4.4
The Large Size and Growth of the Asian Internet Market Lures High-Tech Companies

Large	Users (millions)	Year on Year Growth (%)	Medium	Users (millions)	Year on Year Growth (%)	Small	Users (thousands)	Year on Year Growth (%)
United States	95.4	29	Indonesia	2.0	122	Kazakhstan	100.0	43
Japan	47.1	74	Philippines	2.0	52	Kyrgyz	51.6	50
China	22.5	153	Singapore	1.2	26	Nepal	50.0	43
South Korea	19.0	75	New Zealand	0.8	19	Mongolia	30.0	150
Australia	6.6	18	Vietnam	0.2	100	Brunei	30.0	20
Taiwan	5.7	28	Papua New Guinea	0.1	575	New Caledonia	24.0	100
India	5.0	178	Pakistan	0.1	67	Myanmar	7.0	1,300
Malaysia	3.7	48	Sri Lanka	0.1	87	Laos	6.0	200
Hong Kong	2.6	7	Bangladesh	0.1	100	Cambodia	6.0	50
Thailand	2.3	176	Uzbekistan	0.1	60	Turkmenistan	6.0	200

SOURCE: Hachigian and Wu (2003).
NOTE: Countries with 4,000 users or less are Micronesia, Tajikstan, the Solomon Islands, Bhutan, and the Marshall Islands.

Figure 4.22
Worldwide Wage Rates for Direct Labor Electronics Manufacturing

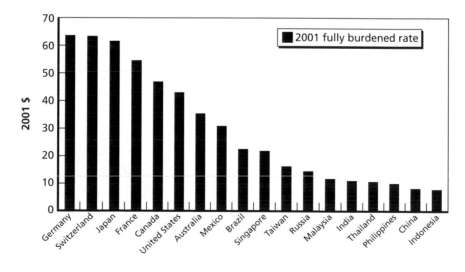

SOURCE: Sherman and Jones (2002).
RAND*TR136-4.22*

Most of the discussion in this report has been focused on the computer hardware and semiconductor industries, the primary focus of the PCAST panel on high-technology manufacturing. The software and IT services industries are undergoing an equally dramatic migration of jobs overseas, mainly because of the differences in wage rates for electronics manufacturing in different countries. The viability of using foreign programmers had been validated earlier when many foreign firms successfully helped patch enterprise and legacy software systems for the millennium Y2K problem. Then, the realization of the economics became apparent. Unlike semiconductor manufacturing, in which costs may be only 5 percent labor, labor accounts for more than 75 percent of the cost of developing software (Amoribieta et al., 2001). This large labor fraction, and availability of lower-cost, skilled programmers overseas, has led much software development to move to India and China. Indian software services are estimated to deliver 60 percent in cost savings to U.S. companies, which account for 70 percent of India's software services exports (Kripalani and Einhorn, 2003). Note that China has not been content with producing software for U.S. and other foreign software companies and is adopting policies designed to build Chinese software developers; one such policy requires all government ministries to buy only locally produced software in an attempt to break dominance of Microsoft on desktop computers (CNETAsia, 2003).

The movement of IT services jobs offshore has been the focus of reports by the Gartner Research Group and the Forrester Research Service. According to Gartner, one out of every 10 jobs within U.S.-based IT vendors and IT service providers will move to emerging markets, as will one out of every 20 IT jobs within user enterprises by the end of 2004 (Morello, 2003). Forrester predicts that the IT services industry will lead an offshore exodus of 3.3 million jobs by the year 2015, including nearly a half-million computer-related jobs (McCarthy, 2002). Some have viewed the offshore movement of software and services jobs as a potential solution to U.S. cost and worker availability issues. In addition to lowering labor costs, outsourcing may alleviate the chronic turnover and worker retention problems for customer service, data entry, and software development jobs (Thomas, 2003).

U.S. and Foreign Market Share vs. Manufacturing Share

Current concerns over U.S. *manufacturing share* do not necessarily translate directly to concerns over the *market share* of U.S. companies. In many instances, leading high-tech companies headquartered in the United States, such as IBM, Intel, Motorola, Dell, Seagate, Cisco, and Solectron, are establishing leading market positions overseas, as they have domestically.

Many U.S. firms have formed joint ventures in China in an effort to quickly establish market and manufacturing positions, resulting in substantial investment in manufacturing facilities. Wholly owned ventures were, on average, more profitable than alliances when foreign investment in China was restricted largely to alliances with struggling state-owned enterprises, but more recent partnerships show more success (Kenevan and Pei, 2003). In 2001, Seagate-China was the leading computer and peripheral equipment company in China, virtually tied with China's Legend Holdings at $2.6 billion. It is interesting to note that, while total production is the same, 90 percent of Seagate-China's production is exported from China, compared with only 18 percent for Legend Holdings. Motorola-China is the second leading manufacturer of communications equipment (including networking, switching, and wireless equipment) with $3.4 billion cost of goods sold in China in 2001, trailing the China Putian Corporation with $5.4 billion. Dell Computers' share of the Chinese PC market has grown from 0 to 10 percent in a short time, which is another example of an emerging U.S.-based firm challenging Chinese domestic suppliers (Sherman and Jones, 2002).

More generally, North American (i.e., U.S. and Canadian) companies have been predicted to continue market leadership in the global information revolution for the foreseeable future (Hundley et al., 2003). But they are in a fierce competition with heavyweight foreign corporations, including Nokia, Siemens, Sony, Samsung, Acer, Ericsson, Legend, and the Taiwan Semiconductor Manufacturing Corporation (TSMC). The competition to reduce costs has led many U.S companies to move manufacturing operations from the Americas and Europe to Asia, where labor costs are lower. The potential of the huge Chinese market has accelerated this rush of corporations to Asia, if establishing a manufacturing presence is viewed as an advantage in establishing market positions.

Characteristics of U.S. Manufacturing vs. Foreign Manufacturing

PCAST and the Office of Science and Technology Policy (OSTP) invited several major high-tech companies to make presentations in order to gain insight into corporate perceptions of the migration of manufacturing overseas. In summarizing these discussions with companies, it is useful to discuss industrial trends and competitiveness along axes commonly used by business consultants, portrayed in Figure 4.23 (Treacy and Wiersema, 1995). U.S. companies are world leaders when competing on *technical excellence*, and most choose to locate high-end R&D and manufacturing in the United States—e.g., when the new product manufacturing process is complex or requires specialized skills, such as the intervention of an R&D engineer.

Figure 4.23
Manufacturing Segmented by Characteristics and Where the United States Generally Dominates or Is Uncompetitive

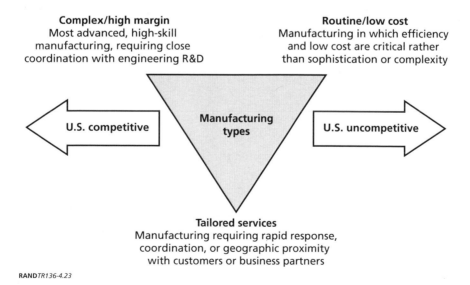

RANDTR136-4.23

When manufacturing processes become "routine," however, U.S. companies that compete on low cost, including those whose core competencies involve *operational excellence*, tend to locate manufacturing operations in lower-cost parts of the world. Advances in supply chain and logistics IT have enabled management and visibility of remote operations. The third axis of business competition frequently discussed along with technical and operational excellence is *client service excellence*, and here the future U.S. leadership is uncertain. Some companies have moved their service operations, such as telephone support, offshore, apparently believing that this type of troubleshooting falls closer to the realm of operational efficiencies. Other firms base their business on understanding intimately their U.S. customers' needs—and responding rapidly with tailored solutions—and may be expected to continue locating their operations in the United States to be near their customers and partners.

One major U.S. networking corporation recounted to PCAST that, a decade ago, all its manufacturing was not only performed in the United States, but it was located in Silicon Valley. With the passing of the North American Free Trade Agreement (NAFTA) in 1993, the company established manufacturing plants in both Mexico and Canada. This experience not only taught headquarters how to remotely manage operations from North America, it also gave it confidence in moving further overseas to Asia. Only two of the corporation's 13 manufacturing centers are now still in San Jose: those operations that need close proximity to engineering staff and those product lines that target the "Buy America" Act. Routine manufacturing that could be performed in lower-cost parts of the world was moved offshore.

U.S. leadership is predicated on riding the leading edge of technology innovation, even as the technologies themselves (and possibly the skills and processes needed to manufacture them) change rapidly. This leadership requires many ingredients, such as world-class universities to research cutting-edge technologies; entrepreneurs and capital support; industries to develop and commercialize innovations; and, of course, consumers able to afford new products. These various parts of our economy have been discussed as a self-stimulating "ecosystem" in which the loss of one may be detrimental to the others. Given the existing health of this ecosystem, U.S. superiority in emerging technology innovations is likely ensured for many years, if not decades.

Summary

The large amounts of data in this chapter have been presented to inform the public debate over the perceived migration of U.S. high-technology manufacturing overseas. The computer, electronics, and semiconductor industries have grown to become among the top five largest manufacturing sectors in the United States, in terms of employment, value added, and shipment value. The IT industry not only propels the U.S. economy through its own growth but has also changed those industries that have adopted IT to grow and reengineer more productive business processes.

Unfortunately, the data on high-tech manufacturing have a serious complication from the effects of a global economic slowdown, immediately following what was an unprecedented dot-com stock market bubble. It is difficult to separate the effects of the recession, from what may have been overoptimistic manufacturing plans during the boom, from the effects of foreign manufacturing competition. As more recent data emerge, closing the gap between measurement and publication dates will make the information more valuable and relevant, as many issues materialize and develop seemingly on "Internet time."

On a very positive note, the most recent data show that U.S. high-tech exports still lead the world by a wide margin, and U.S. high-tech companies have and are expected to maintain leading market share positions for some time. Overall manufacturing output per person has continued to advance, so manufacturing productivity is not in decline. The data paint a picture of an IT industry that, apart from the issue of employment, is actually quite vigorous. The excellence of U.S. research universities, large federal investment in R&D, stable infrastructure, skilled workforce, and other innovation-conducive attributes certainly make it plausible that the United States would remain the leader in emerging technologies for the foreseeable future.

This is not to suggest that challenges do not remain for the United States and its high-tech companies. There has been a decrease in computer manufacturing employment that actually began several years prior to the dot-com crash, roughly a 20 percent decline in computer and electronic products manufacturing employees since at least 1997. U.S. semiconductor manufacturing, which employs as many as the manufacture of computer, storage, terminal, and peripherals sectors combined, also experienced a drop in the manufacturing value added per employee from 1999 to 2001. These trends may be attributed to several different factors, including increases in manufacturing productivity, falling IT product prices, or, stated more generally, the commoditizing of the products.

The rise of foreign high-tech industries is unmistakable. With labor rates in China, India, and elsewhere a small fraction of those in the United States, labor-intensive industries like software development are moving overseas. With some products, like computer peripherals and monitor displays, Asia-based manufacturers are already the leaders. But both peripherals and monitors are the lower value added per employee end of the production scale. The higher value added per employee industries of semiconductor, computer, and storage devices are increasing overseas as well. Additional data are needed to fully understand the relationship between U.S. and overseas manufacturing trends, such as the partnerships between U.S. companies and overseas contract manufacturers.

U.S. manufacturing activities that have remained in the United States tend to be the most advanced, complex manufacturing, typically requiring close coordination with engineering or design staff. But routine manufacturing, in which every efficiency must be pursued, tends to locate overseas. Advances in supply chain IT and other business technologies increasingly make possible asset visibility and operational control of remote operations, further enabling overseas manufacturing for U.S. companies.

Encouraging signs are emerging with the most recent 2003 data on U.S. high-tech industries. U.S. semiconductor and computer industrial production is increasing again, according to the Federal Reserve Bank, even though the communications equipment continues to slide. Computer and electronic products was one of the few areas to show growth from May 2002 to May 2003.

CHAPTER FIVE

U.S. Research and Development Statistics

A major concern regarding a possible loss of U.S. high-technology manufacturing is that it might lead to a reduction in associated research and development if the federal government or industry has less incentive or ability to fund those activities. A decrease in R&D funding might, in turn, lead to a decline in the technology advantage that the United States currently has over other countries.

This section presents R&D funding data from the Department of Commerce, the National Science Foundation, and the recently published RAND/S&TPI report on federal investment in R&D[1] in an attempt to provide preliminary answers to the following questions:

- Has industrial R&D funding varied over time, and, if it has, what caused the change?
- Has federal support for industrial R&D varied over time?
- Has the character of federally funded R&D changed?
- How does U.S. R&D spending compare with that of other industrial countries?

Industrial R&D

Figure 5.1 shows industrial R&D expenditures as a percentage of GDP from 1960 to 1999. Industrial R&D is that performed by for-profit companies and paid for by those companies, the federal government, or other organizations and institutions. While this type is not the only R&D that supports industry, it is a major component.

Over the 40-year period, industrial R&D averaged 1.8 percent of GDP, with a very noticeable dip in the mid-1970s to mid-1980s. In an attempt to understand the cause of the dip, we examine how the portion of industrial R&D funded by the federal government has varied over time.

---

[1] Eiseman et al. (2002). The present document borrows heavily from this report.

Figure 5.1
Industrial R&D Has Varied Between 1.5 Percent and 2 Percent of GDP, with a Low in the Mid-1970s to Mid-1980s

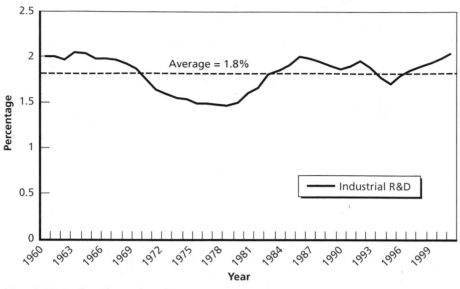

SOURCES: BEA (2003); NSF (2003).
RAND*TR136-5.1*

Figure 5.2 shows the portion of industrial R&D that was funded by the federal government. The difference between the two curves is the portion of industrial R&D called "company-funded." Company-funded R&D is defined for our purpose as the industrial R&D performed within company facilities and funded from all sources except the federal government. The funds are predominantly the company's own but also include those from outside organizations, such as other companies, research institutions, universities and colleges, nonprofit organizations, and state governments. Excluded is company-funded R&D that is not performed within the company as well as company-funded R&D not performed within the United States.

The decline in federal government funding of industrial R&D was the principal reason for the overall industrial R&D decline during the mid-1970s to the mid-1980s. In 1978, federal government funding declined to about 0.5 percent of GDP, compared with 1.2 percent in 1960. If federally funded industrial R&D had remained at the same level of GDP in 1978 as it had been in 1960, total industrial R&D would have been above 2 percent of GDP. Federal government funding continued to decline from 1978 until 1999, when it fell to about 0.25 percent of GDP.

Let us now take a more detailed look at industrial R&D expenditures during the last decade of the 20th century. Figure 5.3 focuses on industrial R&D spending during the 1990s. The average spending was nearly 1.9 percent of GDP, which is above the 40-year long-term average of 1.8 percent of GDP. Thus, despite the nearly total decline in federal government support, industrial R&D funding actually increased over the last decade when compared with earlier decades.

Figure 5.2
A Decline in Federal Funding of R&D Relative to GDP Forced a Dip in Spending—The Decline Has Continued

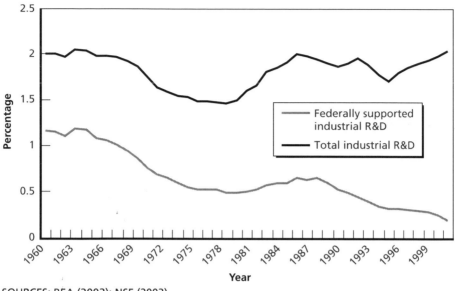

SOURCES: BEA (2003); NSF (2003).

RAND*TR136-5.2*

Figure 5.3
Total Industrial R&D as a Percentage of GDP Generally Increased During the 1990s

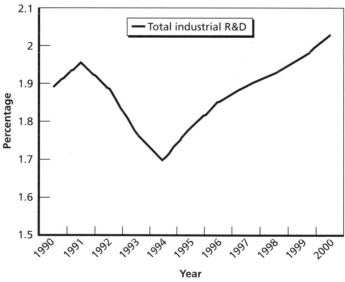

SOURCES: BEA (2003); NSF (2003).

RAND*TR136-5.3*

Figure 5.4 highlights the continuing decline of federal government funding for industrial R&D during the 1990s. In 1990, federal funding was 25 percent of the total. In 2000, the federal government's funding share had declined to 10 percent. Despite the decline in federal government funding, industrial R&D grew because of increased company-funded R&D activities, as noted earlier.

Figure 5.5 compares the 1990–1998[2] trends for company-sponsored[3] R&D performed by all industries, all manufacturers (a subset of all industries), manufacturers of electrical equipment (a subset of all manufacturers), and manufacturers of electronic components (a subset of manufacturers of electrical equipment). The proportion of high-tech industries in each category increases from the industry-at-large category to manufacturers of electronic components, so comparing the trends of the four categories can provide insight into the relative R&D spending by high-tech firms. Spending for each of the four R&D categories is normalized to its 1990 spending level and is set equal to 100. From 1990 to 1998, R&D performed by all industries increased by more than 50 percent. R&D performed by manufacturers of electronic components increased by 175 percent from 1990 to 1998 with a peak of over 250 percent in 1996, relative to the 1990 level. R&D performed by high-tech firms grew much more rapidly than that of all industries. However, there was a steep drop in R&D performed by manufacturers of electronic components from 1996 to 1998. It would be worthwhile to see if that downward trend has continued when more recent data become available. The level of company-sponsored R&D for high-tech manufacturing firms could to be a leading indicator of the future health of those firms.

Figure 5.4
Federally Funded Industrial R&D Continued to Decline Relative to Company-Funded R&D

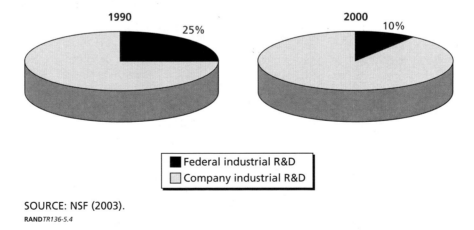

SOURCE: NSF (2003).
RAND*TR136-5.4*

---

[2] The NSF historical data cover 1953 to 1998.

[3] From all sources except the federal government.

Figure 5.5
Selected Trends of Company-Sponsored R&D, 1990–1998

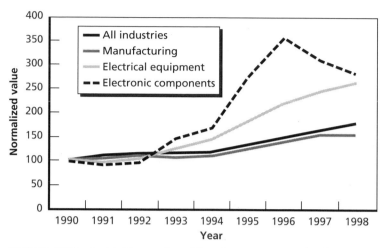

SOURCE: NSF, Industrial Research and Development System (IRIS).

RAND*TR136-5.5*

Federal Funding for R&D

While federal funding for industrial R&D declined over the 1990s—and indeed over the past 40 years—total federal R&D funding has remained roughly the same in constant dollars, as reflected in Figure 5.6.

Figure 5.6
Total Federal R&D Funding Has Remained Nearly the Same in Constant Dollars

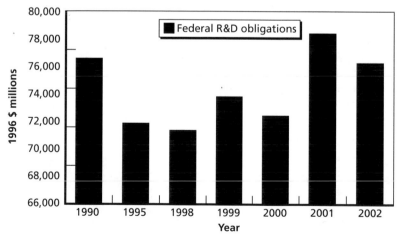

SOURCE: NSF (2002a).

RAND*TR136-5.6*

Figure 5.7 shows that federally funded R&D, as a percentage of GDP, has declined from 1990 to 2002. The decline appears to have recently leveled off at about 0.8 percent of GDP. The decline is a function of GDP growth. If R&D is considered as an investment item, a case can be made that it should keep pace with GDP growth.

The largest federal R&D expenditure has been national defense. Spending on defense R&D has exceeded all other R&D spending (grouped together as "nondefense R&D") for most of the past several decades, although the relative size of the two sectors has varied considerably over the years, as shown in Figure 5.8. Defense and nondefense R&D followed divergent paths during the 1980s and early 1990s. As defense R&D funding increased, nondefense R&D funding decreased, and vice versa. However, after fiscal year (FY) 2000, defense and nondefense R&D funding levels converged, and both increased.

Figure 5.9 shows that there has been a fairly pronounced shift in federal agency funding for R&D over the 1990s. In 1990, DoD provided more than half the federal R&D funds, with the U.S Department of Health and Human Services (HHS) at a distant second. In 2000, the percentage of R&D funded by DoD declined appreciably, while the percentage funded by HHS expanded by about the same percentage as the DoD portion declined.

For defense R&D, the most significant trend has been the dramatic buildup of defense development during the 1980s, followed by a nearly equally dramatic post–Cold War cutback (see Figure 5.10). From FY 1980 to FY 1987, defense R&D nearly doubled, motivated by Cold War tensions with the Soviet Union and emerging technological priorities, such as the Strategic Defense Initiative. However, in the late 1980s and early 1990s, defense R&D fell as the Cold War ended and the "peace dividend" occurred. After bottoming out in FY 1996, defense R&D has been increasing for the past several years and is nearly as large as it was in the mid-1980s.

Figure 5.7
As a Percentage of GDP, Total Federally Funded R&D Has Declined Since 1990—Although the Decline Has Leveled Off

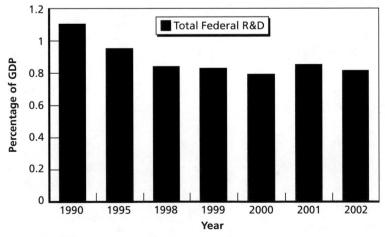

SOURCES: NSF (2002a); BEA (2003).
RAND*TR136-5.7*

Figure 5.8
Trends in Federal R&D Funding for FY 1976–2003 (Budget Authority in Billions of Constant FY 2002 Dollars)

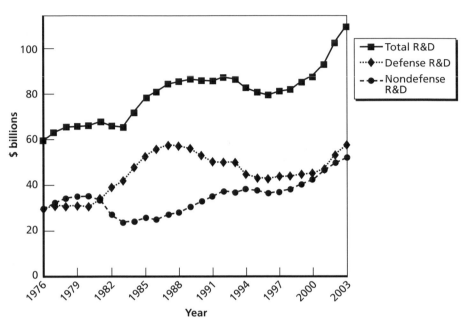

SOURCE: Eiseman, Koizumi, and Fossum (2002).
NOTES: FY 2003 figures are President's request; FY 2002 figures are latest estimates. Includes conduct of R&D and R&D facilities. Constant dollar conversions based on OMB's GDP deflators.
RAND*TR136-5.8*

Figure 5.9
Some Agency R&D Shares Changed, Comparing 1990 with 2002

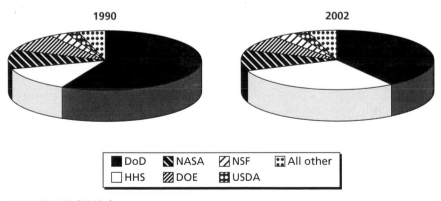

SOURCE: NSF (2002a).
RAND*TR136-5.9*

Figure 5.10 shows that nearly all defense R&D investment is in development, testing, and evaluation of specific weapon systems. This investment fluctuates according to the timing of developmental program starts, the number and expense of individual weapon systems in the development stage, and the relative priority assigned to weapon development within the Pentagon budget.

For nondefense R&D, the past three decades have seen a series of fluctuations reflecting changes in national goals and priorities, as shown in Figure 5.11. The principal long-term trend has been the dramatic increase in health R&D, which now represents the largest single share of the civilian R&D portfolio. A large share of the proposed FY 2003 increase in the funding of health R&D is devoted to bioterrorism research, part of a shift toward the goals of national security and counterterrorism in the FY 2003 budget. "General science" R&D, or science purely for science's sake, is a relatively small part of the U.S. federal R&D portfolio.[4] Although much of the federal R&D investment, and especially the basic research investment, is purportedly for the purpose of laying the knowledge base for future U.S. economic growth, very little of the federal R&D investment is classified as "commerce" R&D (in Figure 5.11, it is one of many missions in the "other" category). Thus, only a small part of the federal R&D portfolio is funded with economic growth as an explicit goal.

Figure 5.10
Trends in Defense R&D: FY 1976–2003 (Budget Authority in Billions of Constant FY 2002 Dollars)

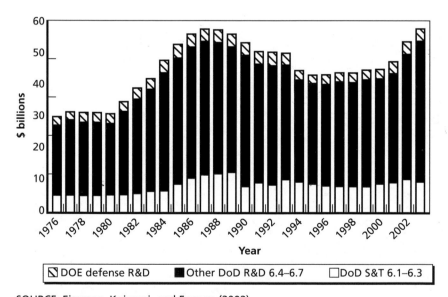

SOURCE: Eiseman, Koizumi, and Fossum (2002).
NOTES: FY 2003 figures are President's request; FY 2002 figures are latest estimates.
DoD S&T figures are not strictly comparable for all years because of changing definitions.
Includes conduct of R&D and R&D facilities. Constant dollar conversions based on OMB's
GDP deflators.
RAND*TR136-5.10*

---

[4] Later, we will contrast U.S. spending in advancement of knowledge research activities with those of other countries.

Figure 5.11
Trends in Federally Funded Nondefense R&D by Function: FY 1976–2003 (Outlays in Billions of Constant FY 2002 Dollars)

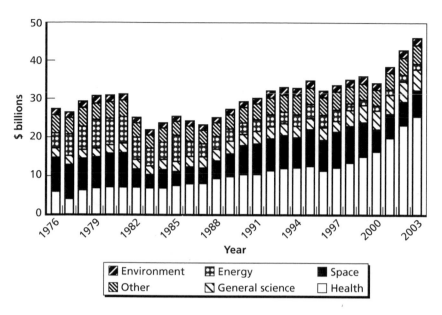

SOURCE: Eiseman, Koizumi, and Fossum (2002).
NOTES: Constant dollar conversions based on OMB's Gross Domestic Product deflators. Excludes R&D facilities. Some energy programs shifted to General Science beginning in FY 1998.
RAND*TR136-5.11*

Figure 5.12 shows the long-term trends of federal R&D funding by government agency. The figure shows both nondefense and defense R&D (it excludes DoD weapon development because of its large size; the DoD line includes the department's science and technology [S&T] only). Again, while fluctuations in agency shares reflect changing national priorities from time to time, the change that stands out is the dramatic increase of the National Institutes of Health's (NIH's) share. The National Science Foundation's (NSF's) budget share has also seen fairly steady growth, albeit at a considerably lower level.

Although the federal government maintains several hundred laboratories, only about one-quarter of federal R&D is conducted at government laboratories (see Figure 5.13). The largest share of federally funded R&D is performed by industrial firms, followed by universities and colleges. The industrial share has fluctuated over time because nearly all DoD development is performed by military contractor firms, and funding for this category has been somewhat volatile because of changing national security priorities. Industrial firms also perform a large portion of DoD "S&T" (i.e., DoD's R&D categories 6.1–6.3) and National Aeronautics and Space Administration (NASA) R&D. A significant amount of R&D is performed under federal grants in universities and colleges, as well as other nonprofit institutions, including federally funded research and development centers operated by contractors, such as the Department of Energy's (DOE's) Argonne National Laboratory in Illinois, run by the University of Chicago.

Figure 5.12
Trends in Federal R&D Funding by Agency, FY 1976–2003 (Budget Authority in Billions of Constant FY 2002 Dollars)

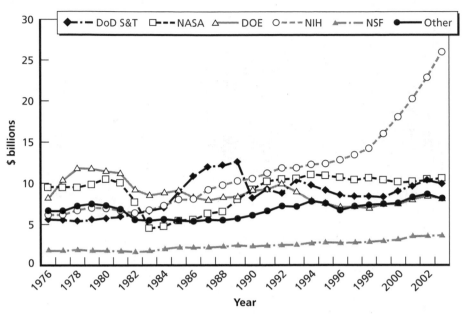

SOURCE: Eiseman, Koizumi, and Fossum (2002).
NOTES: Includes conduct of R&D and R&D facilities. Excludes DoD weapon systems
development. Constant dollar conversions based on OMB's GDP deflators.
RAND*TR136-5.12*

Figure 5.13
Federal R&D Funding by Performer: FY 1976–2002 (Obligations in Billions of Constant FY 2002 Dollars)

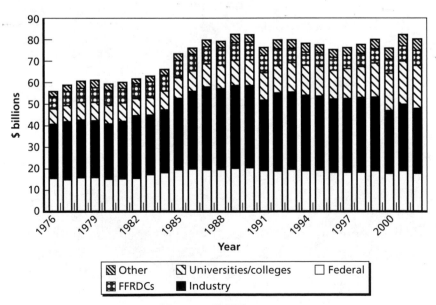

SOURCE: Eiseman, Koizumi, and Fossum (2002).
NOTES: FY 2001 and 2002 data are preliminary. Excludes R&D facilities funding.
RAND*TR136-5.13*

Figure 5.14 shows that a significant increase in federal government R&D funding to academic institutions has occurred over the past decade. From 1991 to 2001, federal funding to academic institutions rose from about $11 billion to nearly $18 billion (in constant 1996 dollars)—a 57 percent increase. GDP in constant 1996 dollars increased 38 percent over the same period.

Thus, the decline in federal funding for industrial R&D appears to be some combination of a decrease in total federal R&D as a percentage of GDP, a change in federal agencies' shares of R&D, and a shift toward more federal support of R&D at academic institutions. That decline, however, has been more than compensated by increased company-funded R&D.[5] Whether the same types of research projects—e.g., basic research versus developmental projects—are being performed is another matter.

Federal Funding for Basic Research

The top half of Figure 5.15 shows the historical trend in overall federal funding for research. Federal obligations for research have increased from about $18 billion in 1970 to about $43

Figure 5.14
A Marked Increase in Federal R&D Funding to Academic Institutions Has Occurred Over the Past Decade

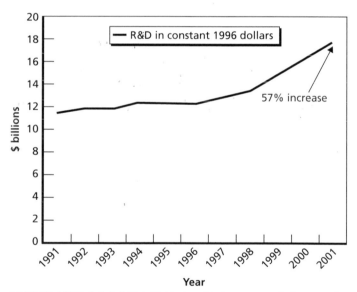

SOURCE: NSF InfoBrief 03-317, National Science Foundation, April 2003.
RANDTR136-5.14

---

[5] There are case studies of federally funded R&D for universities and government labs that led to improved manufacturing processes for next generation semiconductors by Intel and IBM. Thus, even though this was not direct federal funding of industrial R&D, it had the same effect.

Figure 5.15
Federal Obligations for Research, Total and by Broad Field, FY 1976–2002 (Constant FY 2000 Dollars)

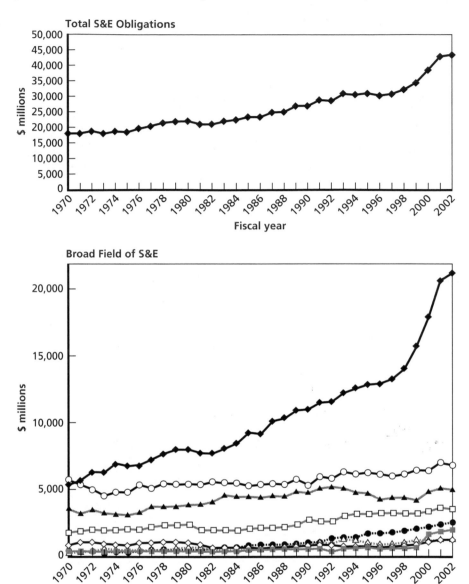

SOURCE: RAND, based on NSF, Federal Funds for Research and Development: Fiscal
Years 2000, 2001, and 2002; and Federal Funds: Detailed Historical Tables: Fiscal Years
1951–2001.
NOTE: FY 2001 and 2002 data are preliminary.
RANDTR136-5.15

billion in 2002 (in FY 2000 dollars)—nearly a 140 percent increase. Perhaps more interesting has been the change in funding by research field.

The bottom half of Figure 5.15 shows the funding trends by broad research area.
While the funding for all fields has generally increased over time, for example, as it has for
math and computer science, there has been a most dramatic increase for life science research,
especially since 1998. The dramatic increase in life science R&D funding raises the issue of
whether R&D funding for other fields should be increased to keep pace, through either the
addition of new funding or a rebalancing of the R&D portfolio.

Twenty-eight federal departments and agencies fund S&E research. Of these, five account for more than 85 percent of the total research funding, and nine account for more than 95 percent of funding (see Figure 5.16). Since 1970, funding of research by HHS has steadily increased, surpassing funding by all other agencies in the early 1970s. In FY 2000, HHS alone funded 47 percent of the total federal S&E research portfolio, as compared with 21 percent in FY 1970. The vast majority of HHS support for S&E research comes from NIH; in FY 2000, NIH accounted for 94 percent of the total HHS-funded research.

The second largest source of research funds is DoD, followed closely by DOE, NASA, and NSF. DOE and NSF have increased their funding of research more than two-and-a-half-fold since FY 1970. In contrast, funding of research by DoD and NASA is only slightly higher than it was in FY 1970; as a share of the total federal research portfolio, DoD-funded research has dropped from 27 percent in FY 1970 to 13 percent in FY 2000, and NASA-funded research has dropped from 21 percent to 10 percent. In FY 2000, DoD, NASA, the Environmental Protection Agency (EPA), and the Department of the Interior (DOI) all funded less research than in FY 1993.

Figure 5.16
Total Federal Obligations for Research, by Agency: FY 1970–2002 (Constant FY 2000 Dollars)

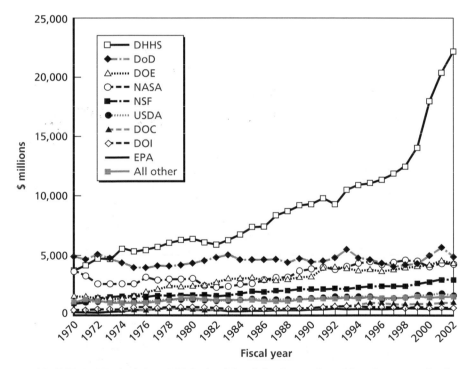

SOURCE: RAND, based on NSF, Federal Funds for Research and Development: Fiscal Years 2000, 2001, and 2002; and Federal Funds: Detailed Historical Tables: Fiscal Years 1951–2001.
NOTE: FY 2001 and 2002 data are preliminary.
RAND*TR136-5.16*

U.S. and Foreign R&D Funding

Figure 5.17 compares total R&D spending for the United States with several individual industrial countries (all are members of the G-7 [Group of 7]) for the period 1981–2001. The measure is billions of 1996 dollars. U.S. R&D spending is considerably larger than that of any other individual country.

The pie charts shown in Figure 5.18 contrast the amount of total R&D spent by the individual countries in 1981 and 2001. Total U.S. R&D is about equal to the sum of the total R&D for the other individual countries shown. The U.S. share of total R&D has grown slightly from 1981 to 2001.

Figure 5.17
U.S. R&D Dwarfs That of Other G-7 Countries

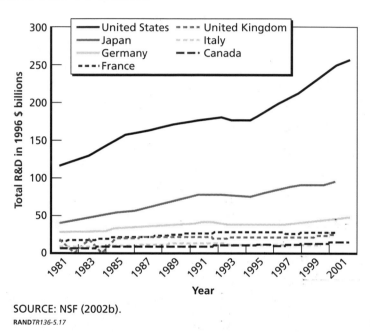

SOURCE: NSF (2002b).
RANDTR136-5.17

Figure 5.18
U.S. R&D Funding Nearly Equals That of Other Countries and Has Grown Slightly

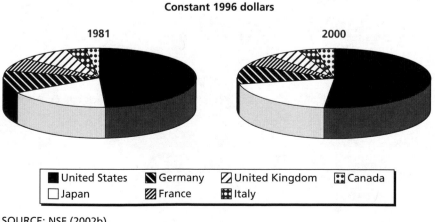

SOURCE: NSF (2002b).
RANDTR136-5.18

Figure 5.19 shows the percentages of R&D expenditures for the G-7 plus Russia by source of funds. In all the countries except for Italy and Russia, industry was the leading source of R&D funds. In contrast, government was the leading source of academic R&D for all eight countries, as shown in the bottom half of the figure.

Figure 5.20 shows the purpose for which the government R&D funds were spent. More than half of U.S. government R&D funds were spent on defense projects. No other country spent anywhere near that percentage on such projects. The United States also spent a considerable portion of its R&D funds on health-related projects, as noted earlier. Note also that the United States spent a much lower percentage of its R&D funds on advancement of knowledge.

Figure 5.21 shows the total R&D spent by the individual G-7 countries as a percentage of each nation's GDP. The United States spends about the same percentage as Japan and Germany.

Figure 5.19
R&D Expenditures by Country and Source of Funds

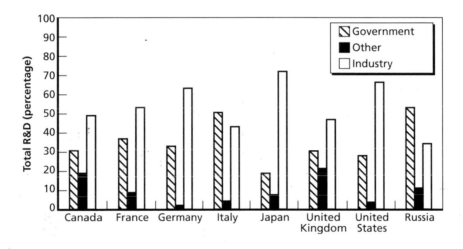

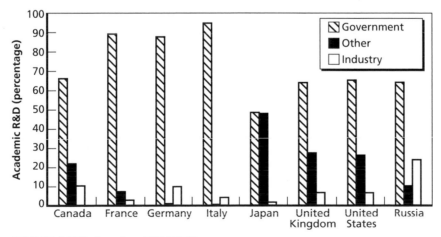

SOURCE: RAND, based on NSB (2002).
RAND*TR136-5.19*

Figure 5.20
Government R&D Support, by Primary Objectives, G-8 Countries

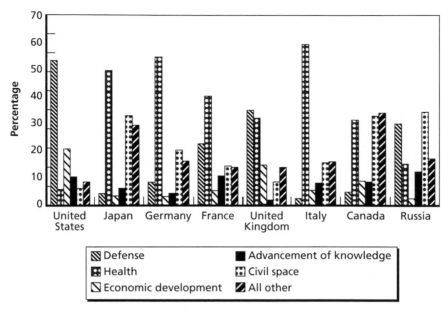

SOURCE: RAND, based on NSB (2001).
NOTE: Data for Italy, Russia, and Canada are for 1998; data for all other countries are for 1999.
RAND*TR136-5.20*

Figure 5.21
As a Percentage of GDP, Total U.S. R&D Is About the Same as That of Japan and Germany

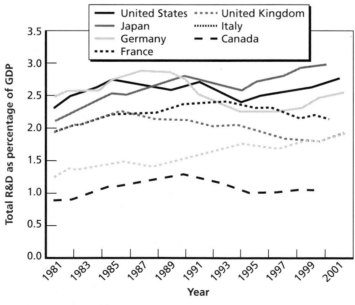

SOURCE: NSF (2002b).
RAND*TR136-5.21*

Table 5.1 shows the ratio of total R&D spending as a percentage of GDP for the United States and 51 other countries. The United States ranks fifth, with Sweden, Finland, Japan, and Switzerland ranking higher. The percentage of U.S. R&D spending is above the average for both the European Union and Organisation for Economic Co-operation and Development (OECD) countries.

Summary

Funding for industrial R&D remained fairly steady over the past three decades and has increased recently. This has occurred despite a decrease in federal funding for industrial R&D. In 1970, federal industrial R&D funding was slightly over 1 percent of GDP. In contrast, by 2000, federal funding had fallen to about 0.25 percent of GDP. Industry has more than made up for the decline in federal support. During most of the 1990s, R&D

Table 5.1
Total R&D as a Percentage of GDP

Country	%	Country	%
Sweden (1999)	3.80	Brazil (1996)	0.91
Finland (1999)	3.22	Spain (1999)	0.89
Japan (1999)	2.93	Cuba (1999)	0.83
Switzerland (1996)	2.73	China (1999)	0.83
United States (1999)	**2.65**	Portugal (1999)	0.76
Israel (1997)	2.54	Poland (1999)	0.75
South Korea (1999)	2.47	South Africa (1998)	0.69
Germany (1999)	2.44	Hungary (1999)	0.69
Iceland (1999)	2.32	Slovak Republic (1999)	0.68
France (1999)	2.19	Greece (1999)	0.68
Denmark (1999)	2.06	Chile (1997)	0.63
Netherlands (1999)	2.05	Turkey (1999)	0.63
Taiwan (1999)	2.05	Argentina (1999)	0.47
Belgium (1999)	1.98	Romania (1999)	0.41
United Kingdom (1999)	1.87	Colombia (1997)	0.41
Singapore (1999)	1.87	Mexico (1999)	0.40
Canada (1999)	1.85	Panama (1998)	0.33
Austria (1999)	1.83	Bolivia (1999)	0.29
Norway (1999)	1.70	Uruguay (1999)	0.26
Slovenia (1999)	1.51	Malaysia (1996)	0.22
Australia (1998)	1.50	Trinidad and Tobago (1997)	0.14
Ireland (1997)	1.39	Nicaragua (1997)	0.13
Czech Republic (1999)	1.25	Ecuador (1998)	0.08
Costa Rica (1996)	1.13	El Salvador (1998)	0.08
New Zealand (1997)	1.13	Peru (1997)	0.06
Italy (1999)	1.03	**European Union (1999)**	**1.86**
Russian Federation (1999)	1.01	**Total OECD (1999)**	**2.21**

SOURCE: RAND, based on OECD (2001) and NSB (2002).
NOTE: Data presented are for latest available year (shown in parentheses). Data for Israel and Taiwan represent these countries' nondefense R&D/GDP ratio.

performed by high-tech firms grew much more rapidly than that performed by all U.S. industries. However, there was a steep drop in R&D performed by manufacturers of electronic components from 1996 to 1998. Whether this trend has continued past 1998 remains to be seen. The level of company-sponsored R&D for high-tech manufacturing firms could be a leading indicator of the future health of those firms.

Over the past decade, total federal R&D funding has remained fairly constant—both in terms of constant dollars and as a percentage of GDP. In 2003, total federal R&D was split nearly evenly between defense and nondefense activities. The major change has been the strong growth in the health R&D component, in which funding has increased about 150 percent since 1990. The dramatic increase in life science basic research raises the issue of whether basic research funding for other fields should be increased to keep pace, through either the addition of new funding or a rebalancing of the basic research portfolio.

In aggregate measures, U.S. total R&D support is comparable to the support of other major industrial nations. In 1981, the United States spent, in constant dollars, nearly as much for R&D as the other G-7 nations combined. In 2000, the U.S. share of total R&D had grown slightly, relative to the total of those same countries. However, the United States spends its R&D funds differently. As a percentage, the United States spends considerably more on defense-related activities and considerably less on advancement of knowledge activities. However, there may be spillover effects from defense-related activities to the advancement of knowledge that this simple categorization does not capture.

CHAPTER SIX

Science and Engineering Degree Statistics

A possible effect of high-technology manufacturing migrating overseas is a reduction in the employment opportunities in high-tech areas. The reduced employment opportunities could include not only job losses in manufacturing but also job losses in related research and development specialties in industry, university, and government laboratories. Reductions of high-tech employment could, in turn, lead to fewer U.S. students majoring in science and engineering fields at higher-education institutions if they believe that suitable employment opportunities will not be available to them upon graduation.

To provide some information on this topic, we examined data assembled by the National Science Foundation.[1] The published NSF data on degrees granted extends only up to 1998–1999. From then until the present time, the global high-tech manufacturing situation has arguably changed more than during any other recent four-to-five-year period. It would be useful to be able to extend the examination of data on degrees granted from 1998–1999 to the present time. However, such data were not available to us. Within the data limitations just described, we address the following three questions:

- Are higher-education students leaving S&E programs to pursue other academic disciplines?
- What are the trends for those academic disciplines closely associated with the IT area?
- How many foreign students did the United States educate, and did the majority of those students plan to return to their homeland or stay in the United States after graduation?

Trends in Science and Engineering Degrees Granted

The first issue we examine is whether the number of degrees granted by U.S. higher-education institutions is keeping pace with the level of the U.S. workforce. As the U.S. workforce increases in size over time, presumably the need for graduates from higher-education institutions generally and for S&E graduates specifically should also increase. Figure 6.1 shows the relative growth of the U.S. workforce, the number of degrees granted at all levels and for all disciplines, and the number of all S&E degrees granted from 1985 to 1998. Each of the data sets is normalized to its value in 1985—i.e., the 1985 result is set

---

[1] This chapter examines data related to the *supply* of students who have earned S&E degrees. Another subject—not investigated here—is what the *demand* is within the U.S. economy for persons with those degrees.

equal to 100. From 1985 to 1989, the U.S. workforce grew at a faster rate than the number of degrees granted in both categories. However, from 1989 to 1993, the number of degrees granted rose at a sharply higher rate than the increase in the size of the U.S. workforce. The increase is such that the normalized number of all degrees granted and the normalized number of S&E degrees granted remain higher than the normalized size of the U.S. workforce. That relationship remains through 1998, the last year for which data are available. Thus, the number of all degrees and S&E degrees granted more than kept pace with the growing size of the U.S. workforce over the period examined.

We now examine the degree data in more detail. Figure 6.2 shows the total number of students (U.S. and foreign students educated in the United States) who received college bachelor's degrees overall (top curve) and the number who received S&E bachelor's degrees. S&E academic fields are the natural sciences, mathematics and computer science, social and behavioral sciences, and engineering. The data in the figure cover the years 1985 to 1998.

Over the period examined, the total number of bachelor's degrees granted climbed from about 1.0 million to 1.2 million. The number of S&E degrees granted kept pace with the relative percentage of S&E degrees granted, staying nearly constant at about 33–34 percent throughout the 14-year period.

Figure 6.3 shows the total number of students who received master's degrees and the number of those who received master's degrees in S&E fields. Over the period shown, the overall number of degrees increased from about 300,000 to 425,000. As shown in the figure, the relative number of S&E master's degrees kept pace with the increase in the overall number of master's degrees granted. The relative percentage of master's degrees in S&E fields remained constant at 22 percent throughout the 14-year period.

Figure 6.1
Relative Trends of the U.S. Workforce Size and the Numbers of Degrees Granted

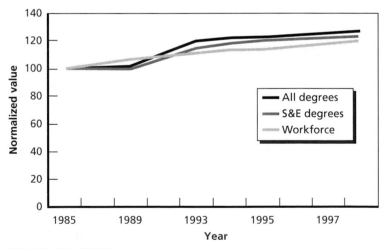

SOURCE: NSB (2002).

RAND*TR136-6.1*

Figure 6.2
From 1985 to 1998, the Percentage of S&E Bachelor's Degrees Granted Remained Nearly Constant

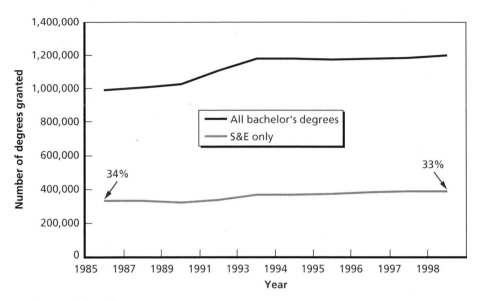

SOURCE: NSB (2002).
RAND*TR136-6.2*

Figure 6.3
The Percentage of S&E Master's Degrees Remained Constant

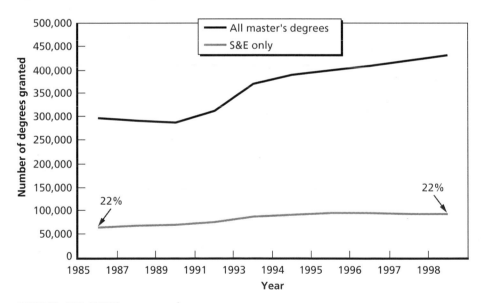

SOURCE: NSB (2002).
RAND*TR136-6.3*

Figure 6.4 shows the total number of students who received doctorate degrees and the number who received doctorate degrees in S&E fields. Over the 14-year period, the total number of degrees granted increased from a little over 30,000 to about 42,000 while the number of S&E doctorate degrees granted increased from about 19,000 to about 27,000. The relative percentage of S&E doctorate degrees granted increased slightly over the 14-year period from 61 percent to 64 percent.

The data on all three levels of S&E degrees are quite consistent, namely, that from 1985 to 1998, the number of S&E degrees granted relative to the number of all degrees granted remained fairly constant. During that period, at least, the relative attractiveness of S&E fields remained constant.[2]

Trends in Degrees Granted for Information Technology Fields

We now turn to an examination of the number of S&E degrees granted in fields that are closely related to the IT area. We focus here on IT because it is the primary area investigated by the PCAST panel on high-technology manufacturing.

For our purposes, we arbitrarily define an IT degree as a degree in mathematics, computer science, or electrical engineering.[3] Using this definition, we show in Figure 6.5 the number of "IT degrees" granted at the bachelor's degree level relative to all S&E bachelor's

Figure 6.4
The Percentage of Those Earning S&E Doctorate Degrees Increased Slightly

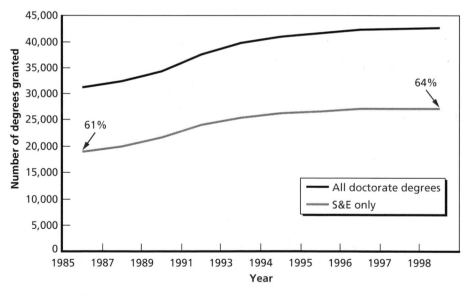

SOURCE: NSB (2002).
RAND*TR136-6.4*

---

[2] It would, of course, be interesting to see if this trend has persisted from 1998 to the present time as data become available.

[3] Chemists, physicists, material scientists, and chemical engineers, among others, are also involved in the IT field.

Figure 6.5
There Has Been a Decline in the Percentage of IT Bachelor's Degrees Granted

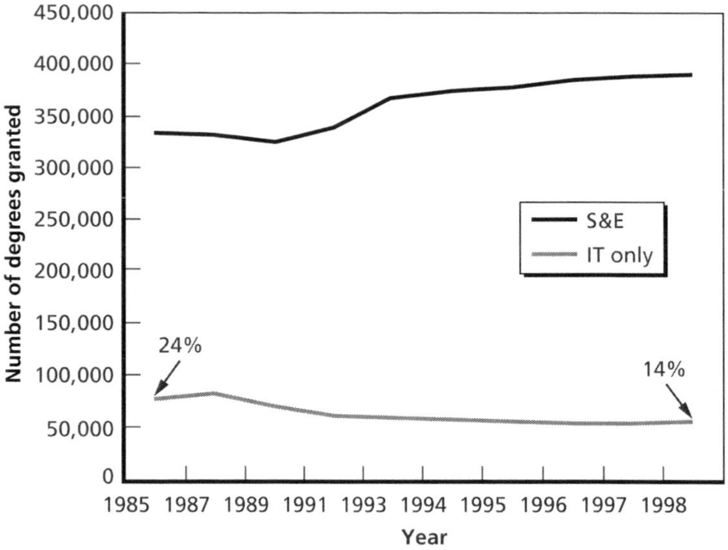

SOURCE: NSB (2002).
RAND*TR136-6.5*

degrees up to 1998. Not all persons who earn a degree in mathematics or electrical engineering will become an IT specialist; however, this is as close as the NSF data will allow us to define an individual who is likely to enter the IT field. The chart shows that the percentage of S&E graduates at the bachelor's level who majored in IT subjects declined from 24 percent to 14 percent over the 14-year period.

We now turn to graduate IT degrees. In contrast to the results shown on the previous chart, Figure 6.6 shows the number of IT degrees granted at the master's level relative to the number of S&E degrees granted increased over the 14-year period from 17 percent to 25 percent.

Figure 6.7 shows that the increasing percentage of IT degrees granted continued at the doctorate level. Over the 14-year period, the relative percentage of IT doctorate degrees increased from 9 percent to 14 percent.

In summary, the data on IT degrees granted appear mixed. Over the period investigated, at the undergraduate level, the relative number of IT degrees granted (using our definition of an IT degree field) declined, but the relative number of graduate IT degrees increased.

Trends in Foreign Students Granted Degrees

We now turn to the issue of the number of foreign students that U.S. colleges and universities are educating and, of those foreign students, the number who plan to return to their homeland or intend to stay in the United States after receiving their degrees.

Figure 6.6
There Has Been an Increase in IT Master's Degrees Granted

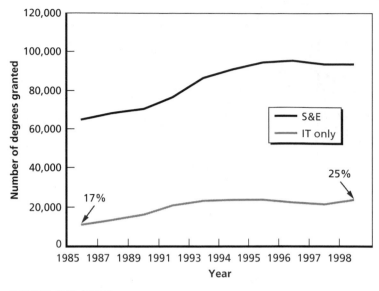

SOURCE: NSB (2002).
RAND*TR136-6.6*

Figure 6.7
The Relative Number of IT Doctorate Degrees Granted Increased

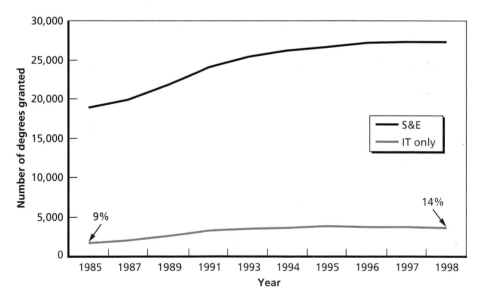

SOURCE: NSB (2002).
RAND*TR136-6.7*

Figure 6.8 shows the foreign graduate student enrollment for either 1998 or 1999 for the United States, the United Kingdom, France, and Japan. We show data separately for S&E minus social and behavioral sciences (SBS) and for computer science and mathematics.

The United States educated a higher percentage of foreign students than the other countries shown, but not much higher compared with the United Kingdom and France. Japan educated a far lower percentage of foreign students in S&E and IT-related fields than the other three countries.

Earlier, we had defined an IT degree as a degree in the field of computer science, mathematics, or electrical engineering. However, as shown in Figure 6.9, the data on foreign students educated in the United States do not provide a breakdown by engineering specialty, such as civil engineering and mechanical engineering. Thus, we can not combine the number of foreign students who receive an electrical engineering degree with the number who receive a degree in computer science or mathematics to get the total number of foreign students who earn an IT degree. As a result, we are limited to showing only the number of foreign students who received a computer science or mathematics degree.

Figure 6.9 shows the number of U.S. citizens plus permanent residents (upper curve) and foreign students (lower curve) awarded an undergraduate degree in mathematics or computer science. The number of foreign students was a very small fraction of the number of U.S. citizens and permanent residents at this education level.

The situation is quite different when we examine the data for the number of master's degrees awarded, as reflected in Figure 6.10. In 1998, the number of foreign students who earned a master's degree in computer science or mathematics was 63 percent of the number of U.S. citizens and permanent residents who earned a degree. This was in sharp contrast to the situation in 1977, when the number of foreign students was 13 percent of the number of U.S. citizens and permanent residents.

Figure 6.8
The United States Had Higher Percentages of Foreign Students in S&E and IT-Related Fields Than the Other Industrial Countries Examined

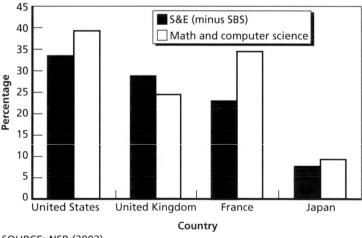

SOURCE: NSB (2002).

RAND*TR136-6.8*

Figure 6.9
Only a Small Fraction of Undergraduate Mathematics and Computer Science Degrees Were Awarded to Foreign Students

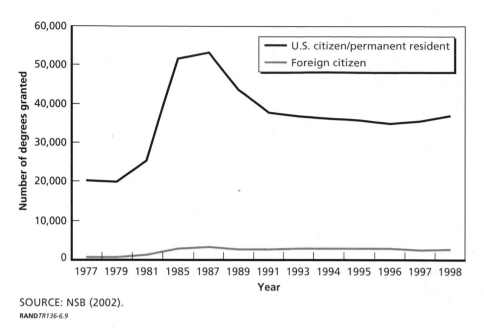

SOURCE: NSB (2002).
RAND*TR136-6.9*

Figure 6.10
A Significant Percentage of Mathematics and Computer Science Master's Degrees Were Granted to Foreign Students

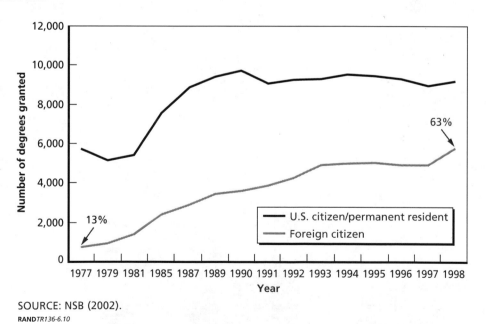

SOURCE: NSB (2002).
RAND*TR136-6.10*

As shown in Figure 6.11, a still higher percentage can be seen when we examine the number of doctorate degrees awarded. In 1999, the number of foreign students awarded degrees was 71 percent of the number of U.S. citizens and permanent residents awarded such degrees.

Figure 6.11
A Significant Percentage of Mathematics and Computer Science Doctorate Degrees Were Granted to Foreign Students

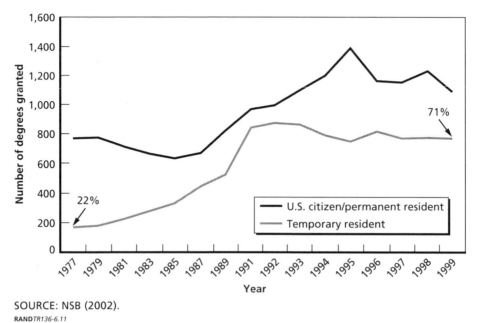

SOURCE: NSB (2002).
RAND*TR136-6.11*

How many of the foreign students who received doctorate degrees intended to stay in the United States after graduation? To answer that question, we show the intention of students from South and East Asia who earned doctorate degrees in the "hard" sciences and engineering fields—i.e., the data exclude degrees awarded in the social and behavioral sciences. We focused on students from South and East Asia because that is the area from which it is believed the severest competition for high-tech manufacturing will come.

The data in Figure 6.12 show that, from 1977 to 1999, the percentage of students intending to stay in the United States increased from 50 percent to over 80 percent. The percentage of students with firm plans to stay in the United States increased from 35 percent to 55 percent over the same period. Thus, it appears from the data that only a minority of foreign students earning a doctorate degree in "hard" science and engineering fields intended to return to their country after being educated at U.S. colleges and universities. What we cannot tell from the data is whether the students' intentions coincided with the actions they subsequently took after receiving their degrees. Also, it would be interesting to see if the trends shown above remained the same or changed in the early 2000s as the United States entered a recession and when employment opportunities may have decreased for graduating foreign students.

Summary

The published NSF data on degrees granted extends only up to 1998–1999. From then until the present time, the global high-technology manufacturing situation has arguably changed

Figure 6.12
A Large Percentage of Asian Students Planned to Stay in the United States After Earning Their Doctorate Degrees

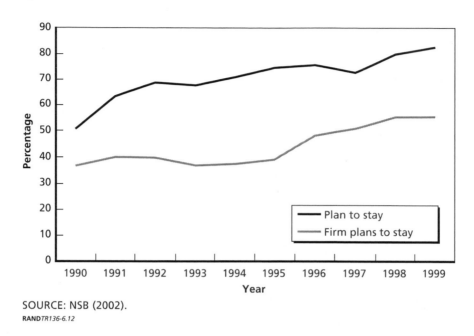

SOURCE: NSB (2002).
RAND*TR136-6.12*

more than during any other recent four-to-five-year period. It would be useful to be able to extend the examination of data on degrees granted from 1998–1999 to the present time. However, such data were not available by the time we completed this study. Within the data limitations just described, we found that:

- The number of all higher-education degrees granted and the number of all S&E degrees granted more than stayed the pace with the growth of the U.S. workforce.
- The number of S&E bachelor's, master's, and doctorate degrees as a percentage of all degrees granted remained remarkably constant from 1985 to 1998.
- The number of bachelor's degrees in computer science, mathematics, and electrical engineering—fields closely associated with the information technology area—as a percentage of all S&E bachelor's degrees declined from 1985 to 1998, but the percentage of graduate degrees in the three fields increased over the same period.
- From a limited sample of industrial countries, we found that the United States educates a larger percentage of foreign graduate students in the fields of computer science and mathematics than do other countries. A significant portion of master's and doctorate degrees awarded at U.S. academic institutions—63 percent and 71 percent, respectively—went to foreign students. However, in 1999, more than 80 percent of Asian students awarded doctorate degrees planned to stay in the United States. The United States needs to consider how it can balance its desire to have foreign graduate students return to their countries, and the potential to rise to leadership positions, with having those same foreign students stay in the United States—especially those who have attained S&E degrees—to contribute to U.S. technological leadership.

- We found no evidence to support the hypothesis that interest in attaining S&E degrees is waning, nor is it apparently being adversely influenced by the growth in foreign high-tech manufacturing.

- In certain IT-related fields, such as math and computer science, the United States is increasingly educating foreign-born students at the graduate degree level. The data show that most foreign students wish to remain in the United States after graduation. Nonetheless, this means that the United States may be increasingly dependent on foreign-born students to fill its IT employment needs at the graduate degree level. Certainly, foreign-born students with advanced degrees in S&E fields should be encouraged to stay in the United States after graduation. But more importantly, programs should be formulated to attract U.S. citizens with S&E bachelor's degrees to seek advanced degrees so that the United States is less reliant on foreign-born students to provide the needed expertise in these areas.

CHAPTER SEVEN

Cross-Strait Information Technology and Investment Flows, and Taipei's Policy Responses

Taiwan is experiencing an outflow of IT production to mainland China. In many respects, the Taiwanese experience mirrors the situation that some fear is happening to the United States, namely, the migration of high-technology manufacturing to overseas locations, in this case, to mainland China. Yet other aspects of the Taiwanese experience are unique and evolve around the particular security situation that Taiwan faces vis-à-vis mainland China. What steps has the Taiwan government taken to stem the flow of its high-tech manufacturing base to China? Have those steps yielded concrete results? Are aspects of Taiwan's experience applicable to the situation facing the United States? This chapter attempts to answer these questions. It describes IT investment flows from Taiwan to mainland China and Taiwan's policies to stem those flows. If Taiwan, with all its political and security concerns, cannot effectively curb the migration of high-tech manufacturing to China, the U.S. government could face even more difficulty and, perhaps, less effectiveness.

It is difficult to overstate Taiwan's significance in global IT production. Taiwanese companies produce approximately 60 percent of the world's notebook computers, 90 percent of its motherboards, 60 percent of its LCDs, 50 percent of its CDTs (computer display terminals), 30 percent of its optical disk drives, and 25 percent of its servers.[1] Until recently, Taiwan was third in the world in IT hardware production value, trailing only the United States and Japan. However, as more Taiwanese manufacturing has relocated to the mainland, China has surpassed Taiwan, becoming the third leading producer of IT hardware, while the island has fallen to fourth. At the same time, the go-go economy in Taiwan that continued throughout most of the 1990s has come to a crashing halt. GDP growth slowed to 3.82 percent during the fourth quarter of 2000 and contracted for the first time in Taiwan's history, declining by 1.58 percent in 2001. The global high-tech bust further encouraged movement to the mainland as Taiwanese companies looked to China to lower production costs and improve competitiveness. China also proved attractive as one of the strongest growth markets worldwide for IT products. The continued flow of investment to the mainland led to charges by some in Taiwan, including many independence supporters, that trade with the mainland was "hollowing out" Taiwan's economy and that restrictions on mainland investment should be maintained. The available evidence suggests, however, that although the changing cross-Strait division of labor presents considerable challenges for Taiwan, the island's economic problems cannot be blamed solely on China's economic magnetism.

---

[1] Interviews with industry analysts at the Market Intelligence Center (MIC), Taipei, Taiwan, October 2002.

In examining these issues, this chapter proceeds in four parts. The first section looks at the development of cross-Strait trade and investment and the evolution of Taiwan's policies toward China-Taiwan economic relations through the late 1990s. The second section assesses more recent trends in cross-Strait trade and investment and Taiwan's policy responses under President Chen Shui-bian. The third section evaluates Taipei's cross-Strait IT policies, with particular attention to the controversy surrounding moves by the island's leading chipmakers to invest in semiconductor manufacturing facilities in China. The fourth and final section of the chapter addresses the contentious debate over whether increasing flows of investment and manufacturing capacity to the mainland are "hollowing out" Taiwan's economy.

The Development of Cross-Strait Trade and Investment and the Evolution of Taiwan's Control Measures

Prior to 1979, there was virtually no economic interaction between China and Taiwan. At the beginning of the 1980s, Taipei enforced a nearly complete ban on exports to the mainland and permitted only certain Chinese foods and medicines to be imported from China via Hong Kong. Despite the prohibitions, Taiwan businessmen rushed to take advantage of increasing opportunities on the mainland, and cross-Strait trade reached nearly $1 billion by 1985. Perhaps recognizing the futility of enforcing the ban on trade and investment, the Taiwanese government in 1985 adopted a noninterference policy on indirect exports to and investment in China.[2] The result was that Hong Kong became the main commercial center for goods shipped to the mainland and the main location for Taiwan-invested subsidiaries or front companies investing in China.

Yet the economy on Taiwan was undergoing important changes that would lead to an accelerated transfer of production to the mainland. Rising wages and the appreciation of the currency reduced the competitiveness of Taiwan's labor-intensive industries and forced these industries to find low-wage production bases, like China. In October 1989, Taiwan issued regulations sanctioning indirect trade with and investment in China. This mix of restrictions and tolerance allowed steady increases in trade between Taiwan and China. In 1978, Taiwan exports to the mainland totaled a mere $51,000, but, by 1991, they exceeded $4.6 billion.

For the Taiwan government, growing economic interaction with China raised concerns that the island was becoming overly dependent on its economic relationship with the mainland and that it was losing its competitiveness. Increasingly, labor-intensive products such as toys, footwear, and textiles, as well as electronic products such as calculators, television sets, tape recorders, and other electrical appliances, were being manufactured on the mainland, causing Taiwan's share of exports in these products to important markets like the United States and Japan to decline. This trend prompted the Taiwanese Ministry of Economic Affairs (MOEA) in October 1990 to issue the "Regulations on Indirect Investment or Technical Cooperation in the Mainland Area," which required firms with investments on the mainland to register the amount and nature of their investment. After April 1991, firms

---

[2] The Taiwan government generally tolerated investment in China as long as projects were relatively small and involved "sunset" industries.

planning to invest more than $1 million on the mainland had to obtain advance approval of their investment, and those investing lesser amounts needed to report their investment to the ministry.

As a result of the new regulations, reporting to the government revealed a total of $750 million in mainland investments. While this number was probably far below actual investments, it did provide the government a better means to track investment. The regulations also legalized Taiwanese investment in some 3,353 products, "mostly in labor-intensive industries such as apparel, footwear, household electronics, and food processing."[3]

In the early 1990s, Taiwanese investment began to move up the manufacturing chain. Migration of IT hardware manufacturing capacity across the Strait was propelled in large part by the requirement to lower production costs. In 1993, Taiwan companies producing PC-related products on the mainland were already benefiting from significant savings over the cost of production on the island, exceeding the savings available in Southeast Asian countries like Malaysia. For example, producing a motherboard in Southeast Asia costs 4 percent less than in Taiwan, but costs 10 percent less in China. Since Taiwan companies producing for export needed to offer competitive prices by world market standards, these differences in manufacturing cost were sufficient to motivate Taiwan companies to invest in China instead of the Association of Southeast Asian Nations (ASEAN) countries (Chung, 1997, pp. 187–188). Thus, the percentage of offshore manufacturing in mainland China increased rapidly from 1990, when Taiwan firms started heading across the Strait in significant numbers, to 1993, when it reached almost 35 percent, surpassing the percentages in Malaysia and Thailand. Offshore production of PC hardware by Taiwanese companies rose from 10 percent in 1992 to 27 percent by 1995 (Chung, 1997, p. 185), when more than 75 percent of switch power supplies and keyboards, about 50 percent of monitors and graphics cards, almost 40 percent of motherboards, and approximately 25 percent of CD-ROM drives and computer mice were made overseas, primarily in China and the ASEAN countries. By the mid-1990s, Taiwan companies were producing more of these items offshore than in Taiwan and were increasingly relocating offshore production to China.

By 1999, about one-third of Taiwan's IT products were manufactured in China, according to Market Intelligence Center estimates. Taiwan's role in the growth of the mainland's IT sector was already so great that one analyst quipped, "It should be called the 'Greater Taiwan Economy' instead of the 'Greater China' one" (Powell, 2000). Taiwanese companies like Acer began shifting production of desktop computers to China in the late 1990s, and, by 2000, nearly one-half of the PCs produced by Taiwanese computer companies were made on the mainland (Lardy, 2002). The migration of productive capacity has continued, with companies pushing the Taiwan government to lift remaining restrictions on high-tech investment. For example, the Taiwan government in November 2001 lifted its restrictions on the production of high-end notebook computers in China.[4] Facing pressure from customers like Compaq, Dell, and Hewlett Packard to cut costs, leading laptop manufacturers like Quanta Computer Inc. started shifting production to the mainland, where notebook PC production costs were about 5 percent lower than in Taiwan (Nystedt, 2001). In 2002, Taiwan notebook computer manufacturers produced about 30 percent of their

---

[3] Sutter (2002b, p. 525). This list was expanded in 1996 to include 4,895 products.

[4] Taiwan is the world's leading producer of notebook PCs. In 2001, companies from the island manufactured more than 13 million notebook PCs, accounting for over 50 percent of the world market.

units in China, up from just over 5 percent in 2001. Industry analysts predict the percentage will climb to 60 percent in 2003. Also increasing rapidly is the number of TFT [thin film transistor]–LCD monitors produced by Taiwanese companies in China. By late 2002, according to the MIC, more than 60 percent of Taiwan's LCD monitor production had moved to China, more than double the 28 percent that was produced there in 2001. Industry analysts forecasted that most of Taiwan's high-end LCD monitor manufacturers would move their production facilities to the mainland by the end of 2003. The move allows them to take advantage of lower production costs and boost their global market share (MIC, 2002).

In all, by the first quarter of 2002, more than 49 percent of Taiwan's IT hardware was made in China, up from about 37 percent total in 2001, according to MIC. The share of IT hardware produced on the island, meanwhile, declined from slightly more than 47 percent in 2001 to about 38 percent in the first quarter of 2002 (Culpan, 2002a). The share of Taiwan's IT hardware production in mainland China is likely to reach 60 percent in 2003, while the portion produced in Taiwan will likely decline to about 26 percent. Taiwanese-invested companies account for the majority of electronic goods exported by China. Indeed, highlighting the extent of Taiwan investment in the Chinese IT sector, MIC industry analysts estimate that Taiwanese-invested companies produce more than 70 percent of the electronics products made in China.[5] Deutsche Bank statistical analysis showing a strong correlation between increases in Taiwanese exports to China and the growth of Chinese electronics exports supports these estimates (Ma, Wenhui, and Kwok, 2002, pp. 5–6).

Taiwan Government Policy Responses Under President Lee Teng-hui

Although the Taiwan government in the early 1990s continued to ease restrictions on cross-Strait economic interaction, there were still lingering suspicions about the political leverage that increased trade and investment would give to Beijing. This prompted President Lee Teng-hui to urge Taiwan investors to divert their investments from China to Southeast Asia. The "Go South" policy, which Taipei initiated in 1994, aimed to prevent the island from becoming overly dependent on its economic relationship with the mainland by encouraging Taiwan companies to invest in Southeast Asian countries. The policy has produced relatively modest results: From 1995 to 2000, Taiwan was the seventh largest investor in the region, with a total of $4.454 billion in FDI.

Having concluded that the "Go South" policy was not reducing the movement of Taiwanese capital to the mainland, the Lee government promulgated the "No Haste, Be Patient" (*jieji yongren*) policy in September 1996, requiring case-by-case approvals for Taiwanese investments in high-tech and infrastructure projects in China. It also placed limits on investments by listed companies and imposed a ceiling of $50 million on individual Taiwanese investments in the mainland. Supporters of the policy argued that it was needed to prevent Taiwan from becoming overly dependent on its economic relationship with China. Some even asserted that Taiwanese investment on the mainland would leave the island vulnerable to economic coercion and enable China to develop military capabilities that would threaten Taiwan's security. But Taipei's policies were beginning to lag badly behind the economic realities of cross-Strait trade and investment.

---

[5] Interviews, Taipei, October 2002.

Recent Trends in Cross-Strait Trade and Investment

Despite the absence of direct links between Taiwan and the mainland, the scope and scale of trade and investment flows across the Taiwan Strait have increased dramatically in recent years, driven in large part by the increasing integration of the Taiwan and mainland China information technology sectors. China's displacement of the United States in late 2001 as the largest market for the island's exports and recent increases in Taiwan investment in the mainland both reflect this important trend.

Cross-Strait Trade

In late 2001, China replaced the United States as Taiwan's largest export market. According to the Taiwan MOEA's Board of Foreign Trade, Taiwan's exports to China in 2001 totaled $24 billion, accounting for 23 percent of the island's total exports, while exports to the United States slipped to 21 percent of the total. During the first half of 2002, Taiwan's exports to the mainland increased by 19 percent year-on-year, while the island's exports to the United States decreased by 9 percent. Economists predict that the growth of Chinese demand for Taiwanese imports "will almost certainly continue to outpace that from other major economies."[6] This is in large part attributable to demand from Taiwanese-invested firms on the mainland.

According to Deutsche Bank estimates, Taiwan's exports to China increased from $4 billion in 1990 to $22 billion in 2001, growing at an average annual rate of 16 percent. Taiwan's imports from the mainland also grew at an impressive rate during the same period, increasing from $800 million in 1990 to $5.9 billion in 2001, which amounts to an average annual growth rate of 20 percent. Moreover, since 1991, China has been the largest source of Taiwan's trade surplus.

Over the past 20 years, China-Taiwan trade has grown from about $10 billion to $30 billion. Now that both China and Taiwan have joined the WTO, the flow of trade across the Strait is likely to grow even more. Nick Lardy predicts that China will become the island's largest trading partner within a few years (Lardy, 2002).

Cross-Strait Investment

The dramatic increase in cross-Strait trade described above has been driven in large part by the expansion of Taiwanese investments in China (Ma, Wenhui, and Kwok, 2002, p. 3). Taiwan firms began investing in China in the 1980s, and Taiwanese investment in China took off in 1987 after the government lifted martial law and began permitting visits to the mainland. After the Taiwan government legalized indirect investment on the mainland in the early 1990s, China quickly surpassed Southeast Asian countries as the destination of choice for Taiwan investors. Initial investments were mainly in labor-intensive industries but later began moving into capital and technology-intensive industries (Chung, 1997, p. 173).

China has become the largest recipient of Taiwanese FDI. Many analysts are calling the current influx of Taiwanese investment in China the "third wave" of investment from the island, and at least one has quipped that "a Tsunami might be a better description" for the

---

[6] Ma, Wenhui, and Kwok (2002, p. 4). The growth of Taiwanese exports to China is a result of two factors: the demand from Taiwanese-invested companies on the mainland for intermediate goods produced on the island and Chinese demand for Taiwanese final products.

current wave of investment flowing across the Taiwan Strait.[7] Taiwanese companies are investing in China to take advantage of lower production costs, especially land, labor, and construction costs; to gain access to the Chinese domestic market, as part of an "investment cluster effect"; and to benefit from preferential tax policies and other incentives offered by the Chinese central government and local governments. Although there is general agreement that these are the primary reasons Taiwanese IT firms are investing on the mainland, there is a wide array of opinion on the relative weight of the factors. For example, according to Deutsche Bank economic analysts, "cost advantage is the primary reason for Taiwanese investments in China." Some Taiwan high-tech executives assert, however, that although lower costs are part of the allure of China for some companies, for many firms, access to the Chinese market and opportunity to sell products in China are much more important.[8] Our assessment is that the latter view most accurately reflects the calculations of many Taiwanese IT companies seeking to expand their investments in China. While lower costs, Chinese government incentives, and increasing availability of skilled engineers and technicians are important factors, the desire to gain greater access to the rapidly expanding China market and the perceived benefits of investment clustering are the primary motivations for many Taiwan firms investing in high-tech projects on the mainland.

While the overall trends in cross-Strait trade and investment are relatively clear, finding solid quantitative data is much more difficult, especially for cross-Strait investment.[9] Much of the money that flows across the Strait does so illegally. Going offshore to circumvent restrictions is an art form that Taiwan investors have developed over the course of 20 years. Yet the information available from official Taiwan government sources only captures approved deals, so it represents only "a fraction of the overall picture," according to a business community source.[10]

Although it is widely believed that the statistics compiled by the MOEA Investment Commission understate greatly the actual amount of Taiwanese investment in China, the official numbers still serve as a basis of comparison with other estimates and as an indication of the increases in approved investment in recent years. From 1991 to 2000, according to MOEA Investment Commission statistics, approved investment by Taiwanese companies in the mainland was $17.1 billion, accounting for about 39 percent of approved overseas investment by Taiwanese firms and making the Chinese mainland the number one recipient of Taiwanese FDI during that period (Republic of China, 2002). According to the Chung-Hua Institute for Economic Research, more than 64 percent of Taiwan's outbound FDI went to China in 2000, up from around 43 percent in 1999 (Lawrence, 2002). The MOEA estimates that Taiwan was the fourth largest investor in China in 2001, following Hong Kong, the United States, and Japan. By early 2002, the Taiwan government estimated that

---

[7] Lawrence (2002). The "first wave" was composed largely of small and medium enterprises that migrated to Guangdong and Fujian in the 1980s to produce toys, textiles, and shoes. Petrochemical companies and lower-end IT hardware manufacturers dominated the "second wave" of Taiwan investment in the mainland, which took place in the early 1990s.

[8] Interviews, Taipei, April 2002 and October 2002.

[9] Analysts at Taiwan's MIC, which is widely regarded as one of the best sources of data on IT in the Greater China region, said it is difficult to gather accurate data on Taiwan investment in China, especially since a significant percentage of it is still illegal and Taiwan companies frequently channel investments through the British Virgin Islands and other countries to circumvent government regulations.

[10] Interview, Washington, D.C., April 2002.

cumulative contracted investment by the island's firms in China had reached about $30 billion.

The most recent Chinese government estimate of total Taiwanese FDI in the mainland is about $52 billion, ranking Taiwan as the third largest source of foreign investment in China. Yet China's statistics also almost certainly undercount Taiwanese investment, as much of it flows through holding companies in the Caribbean. According to Taiwanese statistics, in 1995, investment by Taiwan firms in the British Virgin Islands (BVI) accounted for 15 percent of Taiwan's total outgoing investment. The official number nearly doubled in just five years, reaching almost 30 percent in 2000 and making BVI the second largest recipient of Taiwanese FDI following the mainland. There is ample reason to believe that much of the Taiwanese investment in BVI ultimately finds it way to the mainland. Indeed, along with the rapid expansion of Taiwanese FDI in BVI, the percentage of investment from BVI as a share of total FDI in China rose from less than 1 percent in 1995 to nearly 10 percent in 2000. Another reason Chinese statistics undercount Taiwanese investment is that a large percentage of Hong Kong investment in China actually comes from Taiwanese companies. One Taiwan executive said his best guess is that 50–70 percent of the Hong Kong investment in China is really Taiwanese money.[11]

Thus, according to investment bank economists, available official statistics on cross-Strait economic interaction are "seriously distorted," and "none of the official data sources—whether from mainland China, Taiwan, or Hong Kong—provides a complete description of cross-Strait trade" (Ma, Wenhui, and Kwok, 2002, pp. 2–3). In short, all the official statistics are widely viewed as underestimating Taiwanese investment in the mainland, even by some of Taiwan's government officials. For example, Perng Fai-nan, governor of Taiwan's central bank, estimates that, by the end of 2001, cumulative Taiwan investment on the mainland probably reached $60 billion, more than twice the official Taiwan estimate. Some economists have even estimated that cumulative Taiwanese investment in the mainland may actually exceed $100 billion.[12] Somewhat more conservatively, Deutsche Bank economic analysts estimate that from the early 1990s to the present, Taiwanese companies have invested approximately $50–60 billion in China (Ma, Wenhui, and Kwok, 2002, pp. 9–17). Although we were unable to independently verify the accuracy of these different estimates, based on our interviews with government officials, industry analysts, and corporate executives, we assess that the middle-range figures (estimates of around US$60–70 billion) probably most closely reflect the true scale of Taiwan's investment in the mainland.

Two of the most important trends under way in Taiwan investment in the mainland are the increasing concentration of investment in the electronics sector and accelerating changes in regional investment patterns. Taiwanese investment in the mainland is increasingly flowing into high-tech projects. From 1991 to 2000, Taiwanese companies invested $4.796 billion in the Chinese electronics industry, accounting for about 28 percent of all mainland-bound investments approved by Taiwan's government.[13] In the first five months of 2002, however, more than 49 percent of Taiwanese FDI in the mainland went to the electronics sector, according to MOEA Investment Commission data. The increasing

---

[11] Interview, Taipei, April 2002.

[12] See, for example, Lin (2002).

[13] Cheng (2001).

concentration of investment in the electronics sector is reshaping the composition of Taiwan exports to the mainland, as firms producing IT products in facilities on the mainland require more and more inputs from the island. Statistics on the changing composition of reexports from Taiwan to China via Hong Kong also reflect the shift of Taiwanese investment in the mainland from textiles to information technology products.[14]

The most prominent change in the regional patterns is a shift in the concentration of Taiwanese investment from the Pearl River Delta to the greater Shanghai region. When Taiwan businessmen first began investing in the mainland, the factories they established were concentrated heavily in the provinces close to Hong Kong. Indeed, as of 2000, the Guangdong Province had received approximately 35 percent of cumulative Taiwanese investment in the mainland, ranking it as the first choice of Taiwanese investors, according to official Taiwanese statistics. In second place was Kiangs Province, including Shanghai, which accounted for 34 percent of total Taiwanese FDI in China, and in third place was Fujian Province, with 10 percent of the total.

Over the past several years, however, the interest of Taiwan investors has shifted markedly toward Shanghai, Jiangsu Province, and Zhejiang Province. In 2001, Shanghai and Jiangsu displaced Guangdong as the primary destination for Taiwanese investment on the mainland, accounting for more than 50 percent of actual Taiwanese FDI in China that year while Guangdong's share fell to about 25 percent.[15] Almost 1,100 Taiwanese-invested enterprises were established in Shanghai and Jiangsu Province in 2001, and there are now more than 10,000 Taiwanese companies operating in the area. Despite these gains, Shanghai is still second to Guangdong, where there are more than 14,000 Taiwanese-invested firms (Ma, Wenhui, and Kwok, 2002, pp. 9–11).

The trend toward Shanghai and the Yangtze River Delta continued in 2002, as reflected by Taiwan's MOEA Investment Commission statistics, which show that Shanghai and Jiangsu Province received around 53 percent of approved Taiwan FDI in the mainland in the first eight months of the year.[16] Guangdong Province was the second most popular destination, accounting for about 23 percent of approved mainland-bound investment, and Zhejiang Province was third with slightly more than 9 percent of approved investment.

Taiwan's Policy Responses Under President Chen Shui-bian

The election in 2000 of Democratic Progressive Party (DPP) presidential candidate Chen Shui-bian, who had supported opening the "three links" during his candidacy, raised hopes that restrictions on trade and investment in the mainland would be eased. On January 1, 2001, Chen partially delivered on his pledge by establishing the "mini-three links," which allowed residents of the Taiwan-held offshore islands of Kinmen and Matsu to travel directly to the mainland. The government's policy on the full version of the "three links" has become a highly charged political issue in Taipei. President Chen faces a difficult balancing act. However, he faces pressure from the business community and the Kuomintang and People's

---

[14] See Ma, Wenhui, and Kwok (2002, p. 6).

[15] In 2001, Zhejiang Province was third with roughly 10 percent of the total Taiwanese investment on the mainland.

[16] In all, Jiangsu Province has received about $20 billion in contracted Taiwanese investment, according to official statistics, about $5.4 billion of which has been attracted to Kunshan, a city about an hour west of Shanghai that some call "Taiwan town" or "Little Taipei." Suzhou, for its part, has drawn about $1.4 billion in contracted investment from Taiwan, according to Jiangsu provincial government statistics. For more on Taiwanese investments in the Yangtze River Delta, see Lawrence (2002).

First Party opposition to further ease restrictions on investment and to pursue the establishment of direct links. Conversely, he must manage the demands of some of his own DPP constituents and allies in the Taiwan Solidarity Union (TSU) who fear that increasing economic integration with the mainland will increase unemployment among grassroots DPP supporters and diminish Taiwan's prospects for political independence.

Pressure from the Taiwan business community is an important political consideration for Chen, especially given the state of the economy. The view from the business community is that the government has not given sufficient weight to economic concerns,[17] and this leaves Chen vulnerable to charges from the opposition that he is placing too much emphasis on politics instead of dealing with the economic downturn.

Business community pressure appears to be increasing as a result of the economic downturn in Taiwan.[18] For example, Y. C. Wang, Chairman of Formosa Plastics, has called for the early establishment of direct links on several occasions.[19] Wang warned recently that the competitiveness of Taiwan companies operating in China will decline if China and Taiwan do not move quickly to establish the "three links" ("Businesses to Lose Edge for Lack of Three Links," 2002). Other leading Taiwanese executives warned that the lack of direct cross-Strait links diminishes Taiwan's attractiveness to foreign multinational companies.[20] In addition, President Chen faces pressure from opposition legislators who are urging an accelerated approach to opening direct links.

At the same time, however, President Chen is also facing pressure from independence advocates who are wary that increasing economic integration will narrow the range of political options for the island in the future. Opponents of further opening charge that the establishment of the "three links" would increase Taiwan's dependence on the mainland and enable China to exert greater political, economic, and military leverage against the island. An August 2001 editorial, for example, asserted that the "three links" would "totally undermine Taiwan's national security . . . leaving it fewer and fewer cards to play in its attempt to resist China's ever-mounting pressure and blackmail" ("Dangers of Embracing China," 2001).

TSU legislators say neither the party nor its "spiritual mentor," former President Lee Teng-hui, are categorically opposed to the "three links,"[21] but recent statements suggest the party is hardly enthusiastic about the possibility of Taiwan entering into negotiations with the mainland. For example, TSU politicians argue that the links should be characterized as "state-to-state" (Hsu, 2002), and a TSU spokesman has affirmed that as long as China refuses to renounce the use of force to settle its dispute with Taiwan, establishing the direct links would be equivalent to "relinquishing Taiwan's sovereignty and bringing humiliation to the country" (Low, 2002). The domestic political pressure may be intensified by recent overtures from the mainland concerning negotiations on direct links.

On investment policy, the Chen administration continues to support the "Go South" initiative launched under former President Lee. At the same time, however, the Chen gov-

---

[17] The assertion that "politics shouldn't dominate economic issues" was repeated again and again in interviews with Taiwan high-tech sector executives, conducted in Taipei in April and October 2002.

[18] Indirect travel and shipping increase costs; this is becoming a serious issue for Taiwan companies because margins are "razor thin," as one source put it.

[19] See, for example, "Industry Leader Calls Again for Direct Links with China" (2002).

[20] See, for example, Culpan (2002b).

[21] See, for example, "TSU Not Against PRC Links: Lawmaker" (2002).

ernment has sought to manage the increasing flow of money across the Strait. In May 2001, President Chen announced the formation of the Economic Development Advisory Conference (EDAC) to make recommendations on major economic policy issues (Office of the President of the Republic of China, "Background for Convening the Economic Development Advisory Conference"). Charged with proposing measures to deal with an economic slump that resulted in negative growth rates and record unemployment in Taiwan, the EDAC was composed of prominent businessmen, academics, think tank researchers, and representatives from all of the island's major political parties. Among the 125 members of the EDAC were many of Taiwan's most influential business executives, including Morris Chang, chairman of TSMC; Robert Tsao, chairman of United Microelectronics Corporation (UMC); Barry Lam, chairman of Quanta; Stan Shih, chairman and CEO of Acer; Frank Huang, chairman and CEO of Powerchip Semiconductor; and Wang Yung-ching, chairman of Formosa Plastics.[22] The EDAC was divided into five panels focusing on economic competitiveness, investment, the financial sector, unemployment, and cross-Strait economic and trade relations.

The EDAC announced its 322 consensus recommendations in late August 2001, which President Chen, in a speech at the closing meeting of the conference, pledged to implement.[23] Among its key recommendations, the EDAC's mainland affairs panel urged the government to replace the "No Haste, Be Patient" policy with one of "Active Opening, Effective Management" and to "actively promote" the establishment of direct cross-Strait trade, transportation, and communications links. In its final summary report, the EDAC also recommended that the government allow businesses that had already invested in the mainland without obtaining official permission to "make the appropriate adjustments retroactively" (Republic of China, 2001a).

In terms of specifics, the "Active Opening, Effective Management" policy eliminates the $50 million cap on individual investment projects in China and institutes a simplified review process for mainland-bound investments of less than $20 million.[24] Moreover, the EDAC recommended that "sectors that have no further room for development in Taiwan or which can only survive by investing on the mainland should not be restricted" and those "sectors whose investment on the Chinese mainland may result in a transfer or loss of Taiwan's core technologies should be carefully evaluated" (Republic of China, 2001a). Also as part of the "Active Opening, Effective Management" policy, the Taiwan MOEA in August 2002 implemented new measures that require Taiwanese companies with total investments of $20 million or more on the mainland to submit quarterly reports and annual financial statements detailing their activities on the mainland to the government. The MOEA's Investment Commission also announced plans to further liberalize investment regulations by permitting direct investment in China and reducing the number of items prohibited for Taiwanese companies to invest in on the mainland ("MoEA Implementing New Measures on Investments in Mainland," 2002; "Government to Lift Investment Ban on More Items in Mainland China," 2002). Additional relaxation measures implemented earlier in 2002

---

[22] For a complete list of EDAC members, see Office of the President of the Republic of China, "The List of the EDAC's Members."

[23] For an overview of the EDAC's recommendations in the areas of economic competitiveness, investment, the financial sector, unemployment, and cross-Strait economic and trade relations, see Republic of China (2001b).

[24] For more on the relaxation of restrictions under the policy, see Republic of China (2002).

allowed Taiwanese insurance companies to establish branches in China and permitted Taiwanese banks to make loans to Taiwan-invested companies operating on the mainland. At the same time, however, many in the business community are continuing to press the government to pursue further liberalization of regulations and the opening of direct cross-Strait links.

Taiwan Government Cross-Strait IT Policies

The Taiwanese government has been playing catch-up in terms of its policies on cross-Strait economic relations. For the most part, Taiwan's policies have lagged behind economic trends, often by several years or more.[25] Nowhere has this tendency to fall behind the business curve been more evident than in Taipei's attempts to regulate the flow of investment from the island's companies into the emerging information technology sector on the mainland. Taiwan companies have found innumerable ways over the years to circumvent the restrictions imposed by the Taiwan government, such as incorporating overseas or channeling funds through Hong Kong, the Cayman Islands, and the British Virgin Islands. In recent years, the pattern has continued, with Taiwan modifying regulations to accommodate—and to attempt to shape to the extent possible—emerging trends in the increasingly dynamic China-Taiwan economic relationship.

Even with this apparent change in the Taiwan government's attitude toward regulating cross-Strait economic interaction, most observers expect Taiwan businesses, especially companies in the island's IT sector, to continue to search for ways to skirt regulations while at the same time pressing Taipei for further concessions. In Karen Sutter's words, "To remain globally competitive and to capitalize on commercial opportunities in China, Taiwan businesses will likely continue to pull government policy along while testing and skirting existing restrictions" (Sutter, 2002a). For its part, the Taiwan government appears to be "recognizing the role of government and moving in the right direction," according to a former U.S. government official, but its progress remains "halting and fitful." Indeed, the Taiwanese government's decision in March 2002 to permit Taiwanese firms to invest in semiconductor fabrication facilities on the mainland, provided they fulfill certain conditions, appears to indicate that policy is still largely reactive to unfolding commercial developments, as illustrated by the government's approach to regulating investments in semiconductor manufacturing in China.

Taiwan's Semiconductor Chip Policy

After more than a year of acrimonious political debate and extensive lobbying efforts by major Taiwanese semiconductor companies, including industry leaders TSMC and UMC, the Taiwanese government early in 2002 relaxed its ban on investment in semiconductor facilities on the mainland. On March 29, 2002, Premier Yu Shyi-kun announced the government's decision to allow Taiwanese chipmakers to build 8-inch wafer fabrication plants in China. In the announcement, the Taiwanese government offered the following explanation:

---

[25] Two examples illustrate the reactive nature of Taipei's approach to economic integration with the mainland. First, Taipei did not lift its ban on private-sector cross-Strait exchanges until 1987. Second, the Taiwanese government waited until the early 1990s to legalize investment in the mainland by Taiwanese businesses.

> In view of the future competition for the mainland semiconductor market, the [Taiwanese] government now allows its businesses to invest and develop an early presence there. This is to ensure that Taiwan's businesses can compete effectively with other nations and the Chinese mainland in the future as well as maintain Taiwan's leading international position in the wafer manufacturing industry. However, lifting the ban on investment in the Chinese mainland must follow commitments by Taiwan manufacturers to continuously upgrade the development of the local semiconductor industry and prevent Taiwan's core technologies from flowing into the Chinese mainland.[26]

Taipei-based analysts assert domestic politics was the driving force on the issue of permitting Taiwan companies to open chip foundries on the mainland, with the Chen administration facing intense lobbying by the business community on the one side and heavy pressure from pro-independence politicians on the other. The island's semiconductor companies, led by TSMC and UMC, argued that moving some of their older production equipment to the mainland while they moved ahead with investments in more advanced 12-inch wafer production facilities in Taiwan made good business sense and would contribute to Taiwan's economic recovery. The industry leaders made a concerted effort to persuade the government to relax the restrictions.

On the other side of the debate was the TSU, which strongly opposed the decision to allow Taiwanese semiconductor companies to invest in fabs on the mainland, arguing that the transfer of manufacturing capacity to China would result in increased unemployment and a "hollowing out" of Taiwan's high-tech industry. Other opponents of the measure argued that it would allow China to steal advanced Taiwanese technology and that hollowing out would be accelerated if the relaxation sparked a migration of Taiwanese design and packaging and testing companies to the mainland.[27] Former President Lee Teng-hui suggested that Taipei wait for China to renounce the use of force before permitting Taiwanese semiconductor companies to establish 8-inch fabs on the mainland ("No Rush to Allow Wafer Investment in Mainland China: Ex-President," 2002). Following the announcement of the new policy, Lee publicly criticized the government's decision, warning that China's rapid economic development and Taiwan's increasing dependence on the mainland would threaten the island's sovereignty and national identity.[28] Some TSU politicians even suggested that relaxing the ban would play into a Chinese scheme to absorb Taiwan through economic integration. "They don't want to use weapons to retake Taiwan," a TSU legislator said. "They now want to use economics and money" (Shin-luh, 2002).

As a result of the contentious debate on the island, the long-anticipated liberalization measure was accompanied by the announcement of several conditions.[29] Premier Yu's March 29, 2002, policy announcement stated that:

---

[26] "Premier Yu Shui-kun's Policy Statement on the Liberalization of Mainland-Bound Investment in Silicon Wafer Plants," March 29, 2002.

[27] See, for example, "Plans for Moving Wafer Fabs Brings 'Security Versus Business' Debate" (2002).

[28] "Choose Between Taiwan and Mainland: Lee" (2002). Speaking about the issue of Taiwan's identity at a TSU conference in June, Lee asked, "How can you love the wife and the concubine at the same time?" During the speech Lee also confirmed reports that he had told Premier Yu Shi-kun he would rally 100,000 of his supporters and lead a demonstration to protest the lifting of the investment ban.

[29] Some of the smaller semiconductor manufacturing companies in Taiwan have complained that the conditions provide an unfair advantage to TSMC, UMC, ProMOS Technology, and Powerchip Semiconductor. Those four companies are the

- Taiwanese chipmakers would only be permitted to manufacture 8-inch or smaller wafers and must use mature technologies (0.25-micron and higher process technology) in their mainland fabs.
- Taiwanese chipmakers would be permitted to build an 8-inch plant in China only after they are operating at least one 12-inch fab in Taiwan at full production levels for six months.
- The government would permit the construction of a maximum of three 8-inch fabs on the mainland by 2005.
- The government would only allow Taiwanese companies to transfer used wafer manufacturing equipment to facilities on the mainland (investment in new equipment would be discussed in two years).

Premier Yu also stated that the government would investigate to determine whether any Taiwanese companies made illegal investments in semiconductor foundries on the mainland prior to the lifting of the ban. This will be no easy task for the authorities in Taiwan. Indeed, the government's Mainland Affairs Council (MAC) Vice Chairman Liu Te-shun recently acknowledged that "capital outflows through a third country are not easy to trace." Nevertheless, he vowed that the government's investigations into alleged illegal investments in Chinese chipmakers "will continue very aggressively all the way" (Hung, 2002; Cheng, 2002a). The main target of the investigation is the Semiconductor Manufacturing International Corporation (SMIC). Taiwanese officials found evidence that firms from the island made illegal investments in the newly established Shanghai-based semiconductor manufacturing company, and several Taiwan firms have already agreed to withdraw their investments in the company. The MAC asked others to sign documents declaring that they made no investments in SMIC. The government's investigation is also targeting Grace Semiconductor Manufacturing Corporation and He Jian Technology, although officials have yet to uncover evidence of direct investments from Taiwan in either company.[30] The investigation quickly became a hot political issue in Taipei, with the TSU aggressively publicizing allegations that UMC made an investment in a semiconductor facility in Suzhou before the ban was relaxed.[31] In addition, Premier Yu's announcement stipulated that "in order to ensure that the investment in 8-inch wafer foundries on the Chinese mainland will not have a negative impact on the semiconductor industry and our domestic economy, the government will establish sound and effective management mechanisms."

---

only Taiwanese companies that are starting production of 12-inch wafers and thus will be the only ones eligible to build an 8-inch plant on the mainland under the guidelines announced by Premier Yu.

[30] Taiwanese companies found to have violated the rules will face sanctions, and executives who make false statements will face fines of up to $70,000 and possible imprisonment for as long as five years. Despite the government's efforts, analysts say firms that wanted to make forbidden investments in the Chinese semiconductor industry probably found it easy to circumvent the regulations, as demonstrated by the success with which Taiwan companies invested in notebook PC production on the mainland in defiance of a government ban. Most of the major Taiwanese players were already producing notebooks in China before the ban was eased last year.

[31] Tsai (2002). In June 2002, two TSU legislators, Lo Chih-ming and Lin Chih-lung, disrupted a UMC board meeting, arguing with UMC Chairman Robert Tsao and waving a copy of the *Asian Wall Street Journal* that carried an article about UMC's reported investment on the mainland. For its part, TSMC stated publicly earlier in 2003 that it has adhered to government regulations on investment in the mainland, and senior company executives called on the government to strictly enforce its policies to prevent less scrupulous competitors from gaining an unfair advantage.

In August 2002, the Ministry of Economic Affairs finally announced that it was ready to begin accepting applications from Taiwanese chipmakers interested in building fabs on the mainland, and, in September, TSMC became the first Taiwan semiconductor company to apply to the government for permission to set up a manufacturing plant in China. Another round of political combat will likely ensue if Taiwan semiconductor companies push for further relaxation of the regulations to allow the transfer of more advanced process technology to facilities on the mainland.

The debate over the chip policy also gave rise to calls for technology transfer restrictions and personnel controls. In April 2002, the Taiwan government began drafting a National Technology Protection Law (*guojia keji baohu fa*). The bill was intended to address concerns surrounding the government's decision in late March to permit Taiwanese semiconductor companies to establish 8-inch wafer fabrication facilities on the mainland. Indeed, when Premier Yu announced the lifting of the ban on mainland semiconductor investment, he also pledged that among the government's key objectives was "ensuring that no core technologies, personnel, and capital will be drained-off, and Taiwan's current advantages in high-technology industries will be maintained."[32]

However, even before the March 2002 decision relaxing restrictions on high-tech investment in China, the leaking of sensitive technology to the mainland had become a hot issue in Taiwan politics, in large part as a result of reports concerning the alleged illegal transfer by a TSMC employee of proprietary information to SMIC.[33] In response to the controversy generated by the alleged illegal transfer of TSMC's proprietary information, Taipei established an interagency task force charged with preventing industrial espionage and leaks of sensitive technology to the mainland.[34]

Taiwan's National Science Council (NSC) and Ministry of Economic Affairs asked the Institute for Information Industry's Science and Technology Law Center to help prepare the draft of the proposed science and technology protection law. While preparing to draft the law, researchers studied similar laws in the United States, Japan, and Europe.[35] The draft version of the law categorizes technologies related to national security or gives the island a competitive technological edge as "sensitive technologies." The NSC would have responsibility for determining which technologies are designated as "sensitive" and supervising the enforcement of the law.[36]

Taiwanese businesses seem resigned to the eventual passage of the law, even though many executives say it is unnecessary.[37] They argue that technology transfers restrictions are

---

[32] "Premier Yu Shyi-kun's Policy Statement on the Liberalization of Mainland-Bound Investment in Silicon Wafer Plants," 2000.

[33] The TSMC employee allegedly emailed proprietary information about TSMC's cutting-edge 12-inch wafer manufacturing processes to SMIC on several occasions from November 2000 to January 2001; in February 2001, the employee left TSMC to work for SMIC as a consultant. The prosecutor's office in Hsin-chu has been investigating the case since March 2002.

[34] "Task Force Formed to Check Leaks of High-Technology Secrets: Official" (2002). The task force is composed of representatives from the National Security Bureau and other government agencies.

[35] Parts of the law are modeled after the U.S. Economic Espionage Act, according to NSC officials. See Yu-tzu (2002a).

[36] The National Technology Protection Law is intended to supplement existing laws, including the National Security Law, National Secrets Protection Law, Copyright Law, and Statute Governing Relations Between the People of the Taiwan Area and the Mainland Area.

[37] Interviews, April and October 2002.

not needed because Taiwanese companies already have strong incentives to ensure that their employees do not give away advanced technology to the mainland. The executives say that Taiwanese companies are well aware of the intellectual property rights (IPR) problems multinational companies have encountered in China and that Taiwanese firms will keep higher-end production in Taiwan. Given concerns about the difficulties of protecting intellectual property in China, TSMC and UMC would not immediately build 12-inch fabs in China even if they were allowed to do so, according to one knowledgeable observer.[38] Moreover, the observer said, Taiwan IT companies will avoid setting up high-end R&D centers in China, although they may establish centers that will engage in less sensitive development activities.

The Taiwan government is taking the concerns of the business community into account, but is moving forward with the law, according to Taipei-based analysts. "The regulations should not be too restrictive, but they are still necessary," a senior Taiwanese official said.[39] The official said that technologies created by research institutes that receive government funding, defense technologies, and others designated by the law as "sensitive" need to be protected. The senior official acknowledged, however, that the range of technologies that require protection is difficult to define and will need to be revised from time to time. For this reason, a committee under the NSC will be charged with drafting and periodically updating the list of protected technologies.

Along with the technology transfer restrictions, the Taiwanese government drafted regulations that would require Taiwanese scientists and engineers working in certain high-tech industries to apply for government permission before seeking a job in China. The regulations, which have proven more controversial than the other aspects of the technology protection law, were seen by some analysts as part of an effort to reduce criticism of the relaxation of the policy governing investment in semiconductor facilities on the mainland. Others say the drafting of the regulations on the movement to China of personnel from high-tech companies was also a response to the uproar sparked by media reports indicating that more than 300 engineers fled from TSMC, the world's leading contract manufacturer of semiconductors and the flagship company of Taiwan's high-tech sector, to SMIC. Industry sources in Taipei and Shanghai say the true number of TSMC employees who went to SMIC was actually closer to 100, but the outflow of Taiwanese semiconductor engineers to potential rivals on the mainland is seen by many as a worrisome symbol of the migration of the island's high-tech talent to China.[40]

Some major companies would, in principle, like the government to implement some form of restrictions on engineers going to China. Many Taiwanese companies and multinational corporations with business interest in Taiwan and China, however, are firmly opposed to the regulations that the government has proposed.[41] Executives at foreign high-tech firms

---

[38] Interview, October 2002.

[39] Interview, October 2002.

[40] Sources say most were young engineers who left the company seeking opportunities for rapid career advancement. Similarly, some of them left for stock options and possible IPO (initial public offering) riches. Only one TSMC employee who went to SMIC was a senior manager.

[41] Interviews, April 2002. "I deeply believe the personnel restrictions are wrong," said one executive at a multinational high-tech firm. The most likely result is that China will obtain the same technological knowledge from other sources—China will simply invite engineers from Malaysia, Thailand, or Singapore if Taiwan places restrictions on personnel movements, the executive said.

with operations on both sides of the Taiwan Strait say the regulations would make it difficult to transfer experienced employees to company facilities on the mainland. Executives from TSMC and UMC also expressed skepticism about the proposed restrictions on high-tech personnel movements. Executives at the semiconductor companies told LY (Legislative Yuan, Taiwan's legislature) members that the government should focus on preventing the unauthorized transfer of sensitive technologies rather than attempting to place restrictions on scientists and engineers interested in working on the mainland. The President of the American Chamber of Commerce in Taipei, Richard L. Henson, has also warned that multinational companies will be reluctant to invest in Taiwan if they cannot freely transfer Taiwanese employees to the mainland.

For the Taiwan government, another problem is that restricting the movement of Taiwan citizens is inconsistent with the free and democratic image Taiwan wants to project to the world. Some opponents of the proposed restrictions say that they would constitute an unacceptable infringement of individual liberties and could harm Taiwan's international image (Bradsher, 2002; Yu-tzu, 2002b). Many critics of the plan also say it would likely prove impossible to enforce the restrictions; some commentators suggest the regulations were intended primarily to counter criticism of the relaxation of restrictions on high-tech investments in China.

The ultimate fate of the law has yet to be decided. The draft has been approved by the Executive Yuan and submitted to the LY for review. That review has been a contentious process. President Chen and Premier Yu instructed DPP legislators to make passage of the law in the LY a priority in early 2003, hoping to defuse opposition to the approval of investments in 8-inch wafer fabs on the mainland from the TSU, which has insisted that the investments should not be allowed to proceed until the technology protection law is in place.[42]

The Chen administration is also facing pressure from the business community to permit Chinese nationals to work for high-tech companies in Taiwan. Current regulations prohibit Chinese high-tech specialists from working in Taiwan, although the rules allow them to apply for permission to conduct academic research on the island (Miao-jung, 2002). Taiwan executives say that Chinese engineers are needed to fill jobs left vacant by a shortage of qualified high-tech specialists in Taiwan.

Foreign Government Incentives

China

In China, preferential policies offered by the central government and local authorities in various cities, along with the allure of a fast-growing domestic market, which stands out even more when compared with the sluggish global IT industry as a whole, have proven a potent combination, attracting large foreign investments that are fueling the growth of a nascent domestic semiconductor industry. The increase in fabrication facility construction in China over the past several years has given rise to concerns that Chinese government trade and industrial policies are tipping the scales unfairly in China's favor at a time when the industry

---

[42] See Republic of China (2003); "Chen Instructs DPP Lawmakers to Push Through Technology Bill" (2003); and Hsu (2003).

is struggling worldwide. Indeed, tax subsidies and the desire for market access are the principal draws for many companies planning to invest in semiconductor manufacturing facilities in China.

The most controversial incentive is a value-added tax (VAT) rebate on chips made in China that is offered by the central government. Foreign-made chips are subject to a 17 percent VAT, while chips produced in China receive an 11 percent rebate, effectively lowering the VAT on domestically produced chips to 6 percent. Industry representatives in the United States charge that this tax break constitutes an illegal subsidy.[43] The Chinese government has also offered free land use and other incentives such as tax holidays and reductions to companies building advanced semiconductor manufacturing facilities in China. For example, under the "2+3" incentive plan, which applies to IC manufacturers and software firms, the central government offers a two-year exemption from corporate taxes followed by a 50 percent reduction for the next three years.

Local governments in China are offering their own incentives, frequently attempting to one-up each other in a fierce competition to attract foreign investment in high-tech industries. Nowhere is this regional competition more intense than in the semiconductor industry. Shanghai offers a "5+5" plan that gives chip manufacturers a five-year local tax holiday followed by five years during which they pay 50 percent of the city's normal corporate tax rate; the city also offers reduced interest rates on loans. To compete with Shanghai, Beijing offers grants covering up to 15 percent of total project investment, preferential loan policies, and rent-free land for IC manufacturers; the same benefits are also extended to software companies. Moreover, to draw certain high-tech companies to their city, Beijing authorities reportedly have promised executives that they would beat any offers from Shanghai, an incentive program dubbed "Shanghai +1." The central government and local governments throughout much of China are also actively promoting the establishment and expansion of high-tech industrial parks. In addition, central and local authorities are attempting to expand the country's growing pool of technically proficient labor by offering a broad range of incentives aimed at convincing ethnic Chinese engineers and scientists trained in the United States to return to China. Lax pollution regulations in China are also seen as a hidden subsidy that reduces costs for companies manufacturing in China.

U.S. Senator Lieberman's policy paper charges that the Chinese government's incentive policies "reflect a strategic decision and represent a concerted effort by the Chinese government to capture the benefits of this enabling, high-tech industry," threatening to place China in a position that will allow it to control the price and supply of chips. The Chinese government has indeed embarked on a concerted attempt to boost the Chinese semiconductor industry, and some of its policies, such as the rebate that lowers the VAT on semiconductors produced in China to 6 percent while imported chips are subject to the full 17 percent VAT, may very well constitute unfair trade practices. The judgment that these policies would place China in such a dominant market position, however, seems overblown, especially in light of the relatively slow progress being made by several start-up foundry companies in China, which are leading some industry analysts to revise downward their expectations for the development of China's semiconductor industry. In Shanghai, which is emerging as the center of China's semiconductor industry, construction at one company's fab has

---

[43] See Howell et al. (2003). This report is available online at www.dbtrade.com/publications/books_and_studies.htm (accessed February 2004).

repeatedly fallen behind schedule, and another firm has twice delayed its planned IPO. Moreover, many Chinese companies are now aiming at filling particular niches rather than trying to compete with the established Taiwan companies across the board. As a result, executives from leading semiconductor companies in Taiwan are indicating that establishing manufacturing facilities in China has become a less urgent issue. Moreover, only one company has submitted an application to invest in China, despite the flaring of concerns over the past few years that the island might quickly lose its edge in the foundry business to newly established mainland firms.

Taiwan

Taiwan essentially established the pure-play foundry industry, which today is dominated by two of the island's leading high-tech companies, TSMC and UMC. In 2002, the two foundry giants combined to account for more than 70 percent of the global market for made-to-order chips. Both TSMC and UMC were spun off from a government-funded high-tech research institute. When TSMC built its first fab in 1986, the government of Taiwan contributed nearly one-half of the initial $200 million investment. In addition, the government of Taiwan reportedly offers substantial tax incentives to TSMC and UMC. Incentives designed to encourage foreign companies to establish regional headquarters and R&D centers in Taiwan include two years of free rent on land in designated industrial districts, followed by another four years of reduced rental rates, and a variety of corporate income tax breaks.

To ensure that Taiwan companies do not shift too much capacity or move their most advanced production lines to China, the government of Taiwan allows Taiwanese firms to invest only in facilities using older technology on the mainland and requires that they must invest in state-of-the-art facilities in Taiwan to qualify for permission to invest in China. Even though the concerns about competition from China in the semiconductor industry that motivated those restrictions have decreased recently, the island's chipmakers must still contend with several other problems. Earthquakes and water shortages have created recurring problems for Taiwan's world-leading semiconductor foundry industry. Some analysts expect that concerns over these issues, as well as worries about relations with China and the SARS (Severe Acute Respiratory Syndrome) outbreak earlier this year, will likely lead some customers to diversify by seeking second sources of supply.

Summary

China's displacement of Taiwan as the world's third largest producer of information technology hardware items, in large part as a result of the relocation of so much of the island's IT production capacity to the mainland, has touched off a debate in Taiwan about the supposed "hollowing out" of the island's economy (Lardy, 2002). A process of restructuring, in which manufacturing jobs have been flowing out of Taiwan as the island moves into the production of high-tech products, has been under way for well over a decade. The number of people employed in manufacturing in Taiwan fell by 16 percent between 1987 and 1995 (Naughton, 1997, p. 13). For most of this period, however, incomes rose and unemployment remained low, but the economic downturn that began in 2001 made the loss of manufacturing jobs into a controversial issue in Taiwan politics.

Many politicians and pundits in Taiwan blame the increasing outflow of investment capital to the mainland for the recent economic downturn in Taiwan. Capital flows to China are a scapegoat for low economic growth, sluggish domestic investment, and record levels of unemployment. The supposed "hollowing out" of the island's economy has become a highly charged political issue in Taipei. A report issued by the Control Yuan (the Taiwanese government's highest-level monitoring and auditing body) in September 2002, for example, criticized the government for failing to stem the tide of capital flowing to China. The report asserted that this was one of the major causes of the economic downturn on the island ("Cabinet Failure to Stem Capital Outflow Censured," 2002). Similarly, a recent editorial charged that "China fever has borne no economic fruit for Taiwan over the past decade" and has "diverted resources away from . . . integration with the innovative developed world, investment in basic and applied research and the building of a modern infrastructure conducive to sustained growth in living standards."[44]

The public is also concerned about the supposed "hollowing out" of Taiwan's economy. According to a poll conducted in October 2002 by a think tank affiliated with former President Lee Teng-hui's Taiwan Solidarity Union, which has publicly articulated its support for independence, the widespread view in Taiwan is that the flow of Taiwanese investment to the mainland is the main cause of the economic downturn. For example, more than 66 percent of respondents to the poll indicated that they suspect the relocation of Taiwanese manufacturing to the mainland has led to rising unemployment on the island ("Downturn Due to 'China Fever': Poll," 2002).

However, other observers argue that the concerns over the migration of manufacturing capacity to the mainland are misplaced and that the hollowing-out argument is largely political. "The movement of manufacturing is very natural for industry," one senior Taiwan government official said. "It should be seen as a matter of historical and economic evolution, not a political issue."[45] Echoing the official's views, senior executives at many high-tech companies said they do not believe that Taiwan's economy will be hollowed out.[46] Moreover, most economists and industry analysts cast doubt on the hollowing-out argument, assessing that Taiwan's economic problems are not the result of increasing flows of Taiwanese investment to the mainland. Recent analysis by Deutsche Bank economists, for example, questions the hollowing-out thesis, finding little correlation between investment in the mainland and Taiwan GDP growth, and almost no correlation at all between Taiwan investment in the mainland and manufacturing employment on the island. The report concluded that there is "no convincing evidence that Taiwanese investment in China has actually caused a 'hollowing out' of the Taiwanese economy" (Ma et al., 2002, p. 1) and that "mounting Taiwanese capital outflow to the mainland is not the root cause of Taiwan's economic recession seen in 2001" (Ma et al., 2002, pp. 17–19). Similarly, other economists have argued that structural problems with Taiwan's economy, including government budget deficits, state dominance of the banking sector and an excess of nonperforming loans, and

---

[44] See, for example, Lin (2002).

[45] Interview, Taipei, October 2002.

[46] Interviews, Taipei, April 2002 and October 2002.

the effects of the global economic downturn—especially given Taiwan's dependence on IT exports to the sluggish U.S. economy—are the real culprits.[47]

Some who cast doubt on the hollowing-out thesis argue that Taiwanese investment in China and the shift of capacity to the mainland are part of a naturally evolving division of labor that is allowing Taiwan to outsource production while at the same time upgrading its industrial structure. According to Deutsche Bank, "Taiwanese investment in the mainland has led to a new division of labor across the strait, with the Taiwanese conducting R&D, producing key components, and taking orders on the one side, and the mainland producing/assembling on the other side. In many cases, this division of labor requires heavier investment in Taiwan than in the mainland" (Ma et al., 2002, p. 20). Taiwan's MOEA estimates that the number of Taiwan companies that take orders in Taiwan but manufacture their products overseas rose from 16.6 percent in September 1996 to 22.5 percent in December 2001. According to the MOEA report, about 40 percent of information industry production has moved abroad ("Taiwan Is Worried About Companies Moving Production Abroad," 2002).

The movement of industry to China will be gradual, and the problem Taiwan faces is determining what it will do to increase value added at home. The outcome depends on whether Taiwan replaces what is leaving with more advanced, high-added-value products. "The key is managing the transition," one researcher said. Taiwan must move into more sophisticated production as labor-intensive, environmentally demanding production becomes increasingly outsourced.

Yet this entails many challenges for Taiwan. One official noted that Taiwan companies invest less upstream (cutting-edge R&D, product design, development of their own intellectual property) and downstream (have their own brands, provide services), but stay in the middle (manufacturing) for the most part, which is not a sustainable strategy. "Taiwan can't stay in the middle forever," the official said. "The challenge we face is moving in both directions." Taiwan government incentive programs are focusing on this problem. The plan includes several elements. The first is encouraging global companies to establish R&D centers in Taiwan.[48] The second element is moving up from original equipment manufacturer to original design manufacturer to design and development. Global companies can shift some development and application research to Taiwan. Notebook design, for example, is already done in Taiwan, and cell phone, PDA, and digital camera design will soon follow. This is a change that is helping Taiwan IT companies move upstream, the official said. The plan calls for design work to be undertaken in Taiwan and manufacturing to be done on the mainland.

Another senior official said that global companies would increasingly use Taiwan as a "hub" for their Greater China businesses. Businessmen in Taiwan also envision another potential role for the island: a bridge to the mainland for multinational companies. The island can be a technology, R&D, and information hub. "Taiwan is a place where you can protect company information because it has a well-established rule of law," the official said,

---

[47] See, for example, Heaney (2002). The article quotes Paul Cavey, chief economist at the Economist Intelligence Unit, who says, "The hollowing out to China is not the problem. The 2001 slump was triggered by slow global growth." In addition, Cavey argues that the simultaneous increase in Taiwan investment on the mainland and decrease in the island's GDP is "a situation of coincidence rather than cause and effect."

[48] The Taiwanese government is reportedly planning to offer income tax exemptions to attract foreign technology professionals to Taiwan as part of an effort to boost R&D on the island.

implicitly contrasting the island to China, where IPR violations constitute a major problem for many foreign companies. It is worth noting, however, that IPR problems in Taiwan remain a source of friction in U.S.-Taiwan relations, and Taipei still has much to do in this regard if it is to play such a role for the world's high-tech leaders.

As for Taiwan's own high-tech firms, Taiwanese businessmen say companies that cannot make money if they stay in Taiwan will move their manufacturing operations to China but keep their sensitive technology in Taiwan. The semiconductor industry's advanced R&D and advanced technology, for example, will stay in Taiwan, though some companies may set up less advanced R&D centers in China in the future. For their part, Taiwan high-tech executives plan to keep most of the sector's R&D in Taiwan even as manufacturing shifts to China. It is in the interests of Taiwan companies, they say, to "maintain some balance."

To put the current situation in perspective, the semiconductor industry is one of the last waves of the movement of production to the mainland. The wholesale movement of lower-end technology assembly, such as motherboards, keyboards, and computer mice, preceded the movement of semiconductors. It is important to note, however, that most of the fabless companies moving to China are part of the low-end of the market. The low-end of IC design will move to China, but the high-end players will stay in Taiwan. It is important to note, however, that the U.S. experience with Taiwan suggests that this may not be the case for long and that the high-end Taiwanese players are likely to migrate to China as well. The timing might be even faster in the case of Taiwan for several reasons, including the rapid rise up the learning curve in China, the establishment of "mirror" research groups at Chinese universities by several leading expatriate Chinese who are faculty members at U.S. universities, and the exploding population of doctoral students graduating from Chinese universities.

The lower-end companies are trying to take advantage of a vast pool of engineering talent and lower costs by moving to the mainland.

The Taiwan government also has ambitious plans to help develop the island's prowess in sectors like nanotechnology, biotech, and IC design. Taiwan will invest more than $600 million in developing its nanotechnology industry.[49] With the level of funding devoted to the development of the nanotechnology industry, the government has clearly identified the development of the sector as one of the key elements of its plan to upgrade Taiwan's economy even as more manufacturing operations shift to the mainland. Another area in which Taiwan has high hopes for the future is biotechnology (Cheng, 2002b). The Hsinchu science-based industrial park recently opened a new extension dedicated to attracting biotech companies, and the government plans to spend $850 million on R&D to encourage the expansion of the industry. In addition, the government has announced plans to spend about $200 million over the next six years to support the development of the island's IC design sector.

In October 2002, Taiwan established a new science-based industrial park in Taichung. It will be the island's third science-based industrial park when it opens in early 2004 and will focus on attracting nanotechnology companies. The NSC, which supervises the management of science-based industrial parks, is planning to make central Taiwan a nanotechnology center and to support the development of an optoelectronics cluster in the southern part of the island to complement the semiconductor industry that is centered on

---

[49] See, for example, Ho (2002) and "Nano-Technology Gets Boost from Cabinet" (2002).

the Hsinchu science-based industrial park in the north. It is hoped that these new technological initiatives will allow the island to leap ahead while continuing to exploit China as its production base.

We assess that the hollowing-out debate will remain unresolved in Taiwan for the foreseeable future. At the most basic level, as long as Taiwan's relationship with China remains a politically contentious issue, some people will continue to argue that closer economic integration with the mainland threatens Taiwan's security and undermines its economic prospects. The issue is unlikely to die down in Taiwan politics anytime soon for several more specific reasons: (1) cross-Strait trade and investment has been expanding at a blistering pace in recent years; (2) Taiwan's economy has been stumbling, and unemployment has hit record levels; and (3) a presidential election is coming up in March 2004, and economic issues are expected to be major themes in the campaign for both the incumbent (President Chen Shui-bian of the DPP) and the opposition (presidential challenger Lien Chan of the Kuomintang and his running mate James Soong of the People's First Party).

In many ways the Taiwan experience mirrors the situation that some fear is happening to the United States. It is important to note how ineffective Taiwan's policies appear to be at stemming the tide of relocation of lower-end manufacturing to lower-cost locations and to China in particular. If Taiwan, with all its political and security concerns, cannot effectively curb the location of manufacturing to China, the U.S. government could face even more difficulty and even, perhaps, less effectiveness.

Bibliography

Acs, Zoltan J., David B. Audretsch, and Maryann P. Feldman, "R&D Spillovers and Recipient Firm Size," *Review of Economics and Statistics*, Vol. 76, No. 2, 1994, pp. 336–340.

AFL-CIO—*see* American Federation of Labor–Congress of Industrial Organizations.

AIAA—*see* American Institute of Aeronautics and Astronautics.

AICUM—*see* Association of Independent Colleges and Universities in Massachusetts.

American Federation of Labor–Congress of Industrial Organizations, Industrial Union Council, *Revitalizing American Manufacturing*, 2003.

American Institute of Aeronautics and Astronautics, *A Blueprint for Action*, final report published in conjunction with the AIAA Defense Reform Conference, Washington, D.C., February 14–15, 2001.

Amoribieta, Inigo, Kaushik Bhaumik, Kishore Kanakamedala, and Ajay D. Parkhe, "Programmers Abroad: A Primer on Offshore Software Development," *The McKinsey Quarterly*, No. 2, 2001.

Association of Independent Colleges and Universities in Massachusetts, *Engines of Economic Growth: The Economic Impact of Boston's Eight Research Universities on the Metropolitan Boston Area*, 2003. Online at www.masscolleges.org/Economic/default.asp (accessed February 2004).

Audretsch, D. B., and M. P. Feldman, "R&D Spillovers and the Geography of Innovation and Production," *American Economic Review*, Vol. 86, No. 3, 1996, pp. 630–640.

Augustine, Norman R., *Report of the Defense Science Board Task Force on Defense Semiconductor Dependency*, U.S. Department of Defense, Office of the Under Secretary of Defense for Acquisition, 1987.

BEA—*see* Bureau of Economic Analysis.

Bell, Daniel, *The Coming of Post-Industrial Society: A Venture in Social Forecasting*, New York: Basic Books, 1973.

Bluestone, Barry, and Bennett Harrison, *The Deindustrialization of America*, New York: Basic Books, 1982.

Bonomo, James, Julia Lowell, John Pinder, Katharine Webb, Jessie Saul, Peter Cannon, Jennifer Sloan, and David M. Adamson, *Monitoring and Controlling the International Transfer of Technology*, Santa Monica, Calif.: RAND Corporation, MR-979-OSTP, 1998.

Borrus, Michael, James Millstein, and John Zysman, "U.S.-Japanese Competition in the Semiconductor Industry: A Study in International Trade and Technological Development," Policy Papers in International Affairs, University of California, Institute of International Studies, No. 17, 1982.

Bout, B., "Keeping Taiwan's High-Tech Edge," *The McKinsey Quarterly* [Special Edition: *The Value in Organization*], 2003.

Bradsher, Keith, "Taiwan Is Trying to Limit Its Engineers' Work in China," *New York Times,* April 26, 2002.

Bureau of Economic Analysis, Gross Domestic Product by Industry. Manufacturing FTE numbers available online at www.bea.gov, accessed 2002.

_____, National Income and Product Accounts Tables, U.S. Department of Commerce, June 26, 2003.

"Businesses to Lose Edge for Lack of Three Links," *China Post,* October 25, 2002.

"Cabinet Failure to Stem Capital Outflow Censured: Watchdog Finds Only 1 Percent of China Investment Sent Back to Taiwan," *Taiwan News,* September 19, 2002.

"Cabinet Fears Legislature Will Stall," *Taipei Times,* June 9, 2003.

CBO—*see* Congressional Budget Office.

Chen, Andrew Chun, and Jonathan Woetzel, "Chinese Chips: China Could Soon Become a Major Force in Semiconductors—by Taking a Road of Its Own," *The McKinsey Quarterly,* No. 2, 2002.

"Chen Instructs DPP Lawmakers to Push Through Technology Bill," *China Post,* January 29, 2003.

Cheng, Allen T., "The United States of China: How Business Is Moving Taipei and Beijing Together," *Asiaweek,* July 6, 2001.

_____, "Making China Less Fab-ulous," *Fortune,* September 30, 2002a.

_____, "Taiwan Puts Its Chips on Biotech," *Fortune,* October 28, 2002b.

"Choose Between Taiwan and Mainland: Lee," *China Post,* June 8, 2002.

Chung, Chin, "Division of Labor Across the Taiwan Strait: Macro Overview and Analysis of the Electronics Industry," in Barry Naughton, ed., *The China Circle: Economics and Technology in the PRC, Taiwan, and Hong Kong,* Washington, D.C.: Brookings Institution, 1997.

Cliff, Roger, *The Military Potential of China's Commercial Technology,* Santa Monica, Calif.: RAND Corporation, MR-1292-AF, 2001.

Cline, William R., "Trade and Income Distribution: The Debate and New Evidence," Institute for International Economics Policy Brief 99-7, September 1999.

CNETAsia Staff, "China Blocks Foreign Software," August 18, 2003.

Cohen, Stephen S., and John Zysman, *Manufacturing Matters: The Myth of the Post-Industrial Economy,* New York: Basic Books, 1987.

Cohen, W. M., and D. A. Levinthal, "Innovation and Learning: The Two Faces of R&D," *The Economic Journal,* Vol. 99, No. 397, 1989, pp. 569–596.

Congressional Budget Office, *The Effect of Changes in Labor Markets on the Natural Rate of Unemployment,* April 2002.

Corning Corporation, "Corning Making Major Investment to Expand Its LCD Glass Capacity," press release, November 5, 2002.

Culpan, Tim, "Taiwan IT Storms into Mainland," *South China Morning Post,* April 30, 2002a.

_____, "Acer Laments Taiwan Links Ban," *South China Morning Post,* July 11, 2002b.

Curtis, Philip J., *The Fall of the U.S. Consumer Electronics Industry,* Westport, Conn.: Quorum Books, 1994.

"Dangers of Embracing China," *Taiwan News,* August 28, 2001.

DeLong, Daniel F., "Asia Threatens High-Tech Manufacturing Dominance," *NewsFactor Network,* June 4, 2001. Online at www.newsfactor.com/perl/story/10229.html (accessed January 2004).

DoD—*see* U.S. Department of Defense.

"Downturn Due to 'China Fever': Poll," Central News Agency, October 16, 2002.

Eiseman, Elisa, Kei Koizumi, and Donna Fossum, *Federal Investment in R&D*, Santa Monica, Calif.: RAND Corporation, MR-1639.0-OSTP, September 2002.

Estavao, Marcello, and Saul Lach, "Measuring Temporary Labor Outsourcing in U.S. Manufacturing," National Bureau of Economic Research Working Paper No. 7421, November 1999.

Federal Reserve statistical release G.17 (419), June 17, 2003.

Finan, William F., *The International Transfer of Semiconductor Technology Through U.S.-Based Firms*, New York: National Bureau of Economic Research, 1975.

Flamm, Kenneth, *Mismanaged Trade? Strategic Policy and the Semiconductor Industry*, Washington, D.C.: Brookings Institution Press, 1996.

GAO—*see* U.S. General Accounting Office.

Gay, Daniel, "Slumping Tigers Driven Dragon," *Asiaweek* [Hong Kong], July 27, 2001.

Gourevitch, Peter, Roger E. Bohn, and David McKendrick, "Who Is Us? The Nationality of Production in the Hard Disk Drive Industry," in *The Data Storage Industry Globalization Project Report 97-01*, San Diego, Calif.: University of California, San Diego; Graduate School of International Relations and Pacific Studies, 1997.

"Government to Lift Investment Ban on More Items in Mainland China," *Taiwan Economic News*, August 7, 2002.

Gregory, Gene Adrian, and Akio Etori, "Japanese Technology Today: The Electronic Revolution Continues," *Scientific American,* Vol. 245, No. 4, 1981, pp. J1–J46.

Grossman, Gene M., "Strategic Export Promotion: A Critique," in Paul R. Krugman, ed., *Strategic Trade Policy and the New International Economics*, Cambridge, Mass.: MIT Press, 1986, pp. 47–68.

Grunwald, Joseph, and Kenneth Flamm, *The Global Factory: Foreign Assembly in International Trade,* Washington, D.C.: The Brookings Institution, 1985.

Hachigian, Nina, and Lily Wu, *The Information Revolution in Asia*, Santa Monica, Calif.: RAND Corporation, MR-1719-NIC, 2003.

Hatano, Daryl, "Fab America: Keeping U.S. Leadership in Semiconductor Technology," paper presented at The Future of the U.S. Semiconductor Industry, Washington, D.C., May 8, 2003.

Heaney, Bill, "Economist Says China Can't Be Blamed for Woe," *Taipei Times*, November 7, 2002.

Helpman, Elhanan, and Paul R. Krugman, *Market Structure and Foreign Trade: Increasing Returns, Imperfect Competition, and the International Economy,* Cambridge, Mass.: MIT Press, 1985.

Ho, Laura, "ITRI Addresses Innovation Capabilities," *China Post*, October 31, 2002.

Howell, Thomas R., Brent L. Bartlett, William A. Noellert, and Rachel Howe, *China's Emerging Semiconductor Industry: The Impact of China's Preferential Value-Added Tax on Current Investment Trends*, Dewey Ballantine LLP (prepared for the Semiconductor Industry Association), October 2003.

Hsu, Crystal, "Direct Talks for Direct Links, Government Told," *Taipei Times*, November 2, 2002.

_____, "DPP Promises to Prevent Outflow of Technology to China," *Taipei Times*, February 23, 2003.

Hundley, Richard O., Robert H. Anderson, Tora K. Bikson, and C. Richard Neu, *The Global Course of the Information Revolution: Recurring Themes and Regional Variations*, Santa Monica, Calif.: RAND Corporation, MR-1680-NIC, 2003.

Hung, Faith, "Taiwan's Tech Companies Flout China Investment Restrictions," *Electronic Buyer's News*, August 5, 2002.

"Industry Leader Calls Again for Direct Links with China," *Taiwan Economic News*, April 16, 2002.

Irwin, Douglas A., "Trade Politics and the Semiconductor Industry," National Bureau of Economic Research Working Paper Series No. 4745, 1994.

Irwin, Douglas A., and Peter J. Klenow, "Learning-by-Doing Spillovers in the Semiconductor Industry," *Journal of Political Economy,* Vol. 102, No. 6, 1994, pp. 1200–1227.

Isaac, Randy, "Perspectives on the Semiconductor Industry," paper presented at The Future of the U.S. Semiconductor Industry, Washington, D.C, May 8, 2003.

iSuppli Corporation, Plasma Display Market Tracker, Q1 2003, Stanford Resources.

Jaffe, Adam B., "Technological Opportunity and Spillovers of R&D: Evidence from Firms' Patents, Profits, and Market Value," *American Economic Review*, Vol. 76, No. 5, 1986, pp. 984–1001.

_____, "Universities and Regional Patterns of Commercial Innovation," *REI Review* [Center For Regional Economic Issues, Case-Western Reserve University], 1989.

Joel Popkin and Company, *Securing America's Future: The Case for a Strong Manufacturing Base,* Washington, D.C.: National Association of Manufacturers, June 2003.

Johnson, Chalmers, *MITI and the Japanese Miracle: The Growth of Industrial Policy, 1925–1975,* Stanford, Calif.: Stanford University Press, 1982.

Johnson, George, and Matthew J. Slaughter, "The Effects of Growing International Trade on the U.S. Labor Market," in Alan B. Krueger and Robert M. Solow, eds., *The Roaring Nineties: Can Full Employment Be Sustained?* New York: The Russell Sage Foundation and The Century Foundation, 2001, pp. 260–306.

Jorgenson, Dale W., "Information Technology and the U.S. Economy," *American Economic Review,* Vol. 91, No. 1, March 2001, pp. 1–32.

Jorgenson, Dale W., Mun S. Ho, and Kevin J. Stiroh, "Lessons from the U.S. Growth Resurgence," *Journal of Policy Modeling,* Vol. 25, 2003, pp. 453–470.

Kenevan, Peter A., and Xi Pei, "China Partners: As China's Industries Become More Open to Foreign Investment, Alliances with Local Companies Represent an Attractive and Profitable Option for Many Global Corporations," *The McKinsey Quarterly*, No. 3, 2003.

Kimura, Yui, *The Japanese Semiconductor Industry: Structure, Competitive Strategies, and Performance,* Greenwich, Conn.: JAI Press, 1988.

Kletzer, Lori G., *Job Loss from Imports: Measuring the Cost,* Washington, D.C.: Institute for International Economics, 2001.

Kripalani, Manjeet, and Bruce Einhorn (with Paul Magnusson), "A New Battle Over Offshore Outsourcing," *BusinessWeek Online*, June 6, 2003.

Krugman, Paul R., "Introduction: New Thinking About Trade Policy," in Paul R. Krugman, ed., *Strategic Trade Policy and the New International Economics,* Cambridge, Mass.: MIT Press, 1986, pp. 1–22.

_____, ed., *Strategic Trade Policy and the New International Economics*, 5th edition, Cambridge, Mass.: MIT Press, 1992.

_____, "Economic Shuttle Diplomacy: A Review of Laura D'Andrea Tyson's *Who's Bashing Whom?*" in Paul R. Krugman, ed., *Pop Internationalism,* Cambridge, Mass.: MIT Press, 1996a, pp. 105–116.

_____, "Domestic Distortions and the Deindustrialization Hypothesis," National Bureau of Economic Research Working Paper No. 5473, March 1996b.

Krugman, Paul R., and Robert Z. Lawrence, "Trade, Jobs, and Wages," *Scientific American,* April 1994, pp. 22–27 [Reprinted in Paul Krugman, *Pop Internationalism,* Cambridge, Mass.: MIT Press, 1996, pp. 35–48].

Kuttner, Robert, *The End of Laissez-Faire*, New York: Basic Books, 1991.

Kwan, Chi Hung, "Is FDI in Japan Hollowing Out Japan's Industry?" *China in Transition*, November 8, 2002.

Lake, Arthur, *Transnational Activity and Market Entry in the Semiconductor Industry*, New York: National Bureau of Economic Research, 1976.

Lalonde, René, and Danielle Lacavalier, "The U.S. Miracle," Bank for International Settlements paper, April 18, 2001.

Lardy, Nicholas, "China an Awakened Giant," *The Nation Asia News Network*, October 19, 2002.

Lawrence, Anthony, "Hi-Tech's Promised Land," *Topics Online Magazine* [American Chamber of Commerce, Taipei], November 1, 2002.

Lawrence, Robert Z., and Matthew J. Slaughter, "International Trade and American Wages in the 1980s: Giant Sucking Sound or Small Hiccup?" *Brookings Papers on Economic Activity: Microeconomics 2*, 1993, pp. 161–210.

Lieberman, Joseph I., "National Security Aspects of the Global Migration of the U.S. Semiconductor Industry," white paper, U.S. Senate, Armed Services Committee, 2003.

Lin, Hwan C., "Taiwan Has Overinvested in China," *Taipei Times*, October 31, 2002.

Link, A. N., and J. Rees, "Firm Size, University Based Research and the Returns to R&D," *Small Business Economics*, No. 2, 1990, pp. 25–32.

Lorell, Mark, Julia Lowell, Michael Kennedy, and Hugh P. Levaux, *Cheaper, Faster, Better? Commercial Approaches to Weapons Acquisition*, Santa Monica, Calif.: RAND Corporation, MR-1147-AF, 2000.

Lorell, Mark A., Julia Lowell, Richard M. Moore, Victoria Greenfield, and Katia Vlachos, *Going Global? U.S. Government Policy and the Defense Aerospace Industry*, Santa Monica, Calif.: RAND Corporation, MR-1537-AF, 2002.

Low, Stephanie, "Taiwan's NSB Says Direct Flights Endanger Security," *Taipei Times*, October 30, 2002.

Ma, Jun, Zhu Wenhui, and Alan Kwok, *China-Taiwan Economic Integration: Trends and Implications*, Deutsche Bank, Asia-Pacific Equity Research, September 2002.

Macher, Jeffrey T., David C. Mowery, and David A. Hodges, "Semiconductors," in David C. Mowery, ed., *US Industry in 2000: Studies in Competitive Performance*, Washington, D.C.: National Academies Press, 1999, pp. 245–285.

Market Intelligence Center, "Promotional Strategies and Movement to China to Boost Taiwanese LCD Monitor Production in 2H 2002," Taipei, Taiwan, September 12, 2002.

McCarthy, John C. (with Amy Dash, Heather Liddell, Christine Ferrusi Ross, and Bruce D. Temkin), "3.3 Million US Services Jobs to Go Offshore," research brief, Forrester, November 2002.

McKendrick, David, "Hard Disk Drives," in David C. Mowery, ed., *U.S. Industry in 2000: Studies in Competitive Performance*, Washington, D.C.: National Academies Press, 1999, pp. 287–328.

Miao-jung, Lin, "Taiwan's MAC Reviewing Regulations for High-Technology Workers Going to China," *Taipei Times*, May 8, 2002.

MIC—*see* Market Intelligence Center.

Ministry of Foreign Trade and Economic Cooperation, Almanac of China's Foreign Economic Relations and Trade, issues February 1991–July 1996 for 1990–1995 data.

"MoEA Implementing New Measures on Investments in Mainland," *Taiwan Economic News*, July 31, 2002.

Morello, Diane, "U.S. Offshore Outsourcing: Structural Changes, Big Impact," research note, Gartner, COM-20-4837, July 15, 2003.

NAICS—*see* North American Industry Classification System.

"Nano-Technology Gets Boost from Cabinet," *Taiwan News*, November 6, 2002.

National Research Council, *Dispelling the Manufacturing Myth: American Factories Can Compete in the Global Marketplace*, Washington, D.C.: National Academies Press, 1992.

_____, *Securing the Future: Regional and National Programs to Support the Semiconductor Industry*, Washington, D.C.: National Academies Press, 2003.

National Science Board, *Science & Engineering Indicators: 2001*, Arlington, Va.: National Science Foundation, 2001.

_____, *Science & Engineering Indicators: 2002*, Arlington, Va.: National Science Foundation, 2002.

National Science Foundation, *Changing Composition of Federal Funding for Research and Development and R&D Plant Since 1990*, Arlington, Va., NSF 02-315, April 2002a.

_____, *National Patterns of R&D Resources*, Arlington, Va., October 2002b.

_____, *Research and Development in Industry: 2000*, Arlington, Va., NSF 03-318, May 2003.

Naughton, Barry, ed., *The China Circle: Economics and Technology in the PRC, Taiwan, and Hong Kong*, Washington, D.C.: Brookings, 1997.

"No Rush to Allow Wafer Investment in Mainland China: Ex-President," *Central News Agency*, March 10, 2002.

North American Industry Classification System, *Updates for 2007 Federal Register*, December 27, 2002.

NRC—*see* National Research Council.

NSB—*see* National Science Board.

NSF—*see* National Science Foundation.

Nystedt, Dan, "Quanta to Boost China Investments," *Taipei Times*, November 27, 2001.

OECD—*see* Organisation for Economic Co-operation and Development.

Office of the President of the Republic of China, "Background for Convening the Economic Development Advisory Conference," undated. See www.president.gov.tw/2_special/economic/e_index.html (accessed February 2004).

_____, "The List of the EDAC's Members," undated. See www.president.gov.tw/2_special/economic/e_member.html (accessed February 2004).

Office of Technology Assessment, *U.S. Industrial Competitiveness: A Comparison of Steel, Electronics and Automobiles*, U.S. Congress, July 1981.

_____, *International Competitiveness in Electronics*, U.S. Congress, OTA-ISC-200, November 1983.

_____, *Paying the Bill: Manufacturing and America's Trade Deficit*, U.S. Congress, OTA-ITE-390, June 1988.

_____, *Making Things Better: Competing in Manufacturing*, U.S. Congress, OTA-ITE-443, February 1990.

_____, *Competing Economies: America, Europe, and the Pacific Rim*, U.S. Congress, OTA-ITE-498, October 1991.

Ohmae, Kenichi, *The End of the Nation State: The Rise of Regional Economies*, New York: The Free Press, 1995.

Organisation for Economic Co-operation and Development, *Main Science and Technology Indicators: 2001*, 2nd edition, Paris, 2001.

OTA—*see* Office of Technology Assessment.

Phillips, Heather Fleming, "'Buy American' Rules Problematic for Tech," *San Jose Mercury News*, July 7, 2003.

"Plans for Moving Wafer Fabs Brings 'Security Versus Business' Debate," *Taipei Journal*, March 15, 2002.

Powell, Bonnie Azab, "Taiwan Unleashed: Taiwan's High-Technology Economy Rushes in Where Its New Government Fears to Tread," *Red Herring*, October 2000.

"Premier Yu Shui-kun's Policy Statement on the Liberalization of Mainland-Bound Investment in Silicon Wafer Plants," March 29, 2000.

Prestowitz, Clyde V., Jr., *Trading Places: How We Allowed Japan to Take the Lead*, New York: Basic Books, 1988.

Republic of China (Taiwan), Mainland Affairs Division, Economic Development Advisory Conference, *Final Summary Report, Mainland Affairs Division*, Government Information Office, August 26, 2001a.

_____, "President Chen Shui-bian's Address at the Closing Ceremony of EDAC," Government Information Office, August 26, 2001b.

_____, "Economic Ties with the Chinese Mainland," in *The Republic of China Yearbook—Taiwan 2002*, Government Information Office, 2002. Available online at www.gio.gov.tw.

_____, "Excerpts from Q&A Session of Premier Yu Shyi-kun's Anniversary Press Conference," Government Information Office, January 27, 2003.

Ricardo, David, *The Principles of Political Economy and Taxation*, London: John Murray, Albemarle-Street, 1817.

Rowthorn, Robert, and Ramana Ramaswamy, "Deindustrialization: Causes and Implications," International Monetary Fund Working Paper, WP/97/42, April 1997.

Scott, Robert E., "Fast Track to Lost Jobs: Trade Deficits and Manufacturing Decline Are the Legacies of NAFTA and the WTO," briefing, Economic Policy Institute, October 2001.

Semiconductor Industry Association, "World Market Shares: 1982–1990," table, 2002a. Online at www.sia-online.org/downloads/market_shares_82-90.pdf (accessed February 2004).

_____, "World Market Shares: 1991–2001," table, 2002b. Online at www.sia-online.org/downloads/market_shares_91-present.pdf (accessed February 2004).

Sherman, Randall, and Clive Jones, *Electronics Manufacturing in China*, San Jose, Calif.: Electronic Trend Publications, 2002.

Sherwin, Elton B., Jr., *The Silicon Valley Way: Discover the Secret of America's Fastest Growing Companies*, Rocklin, Calif.: Prima Publishing, 1998.

Shin-luh, Shu, "Strait Just Wafer-Thin," *South China Morning Post*, March 27, 2002.

SIA—*see* Semiconductor Industry Association.

Siekman, Philip, and Laurie Windham, "The Big Myth About U.S. Manufacturing," *Fortune*, October 29, 2001.

Smyth, Joseph S., "The Impact of the Buy American Act on Program Managers," *Acquisition Review Quarterly*, Summer 1999.

Southern Growth Policies Board, *Innovation U.: New University Roles in a Knowledge Economy*, Research Triangle Park, N.C., 2002.

Staelin, David H., chair, *The Decline of U.S. Consumer Electronics Manufacturing: History, Hypotheses, and Remedies*, Cambridge, Mass.: Commission Working Group on the Consumer Electronics Industries, MIT Commission on Industrial Productivity, 1988.

State Statistical Bureau (SSB), using MOFTEC figures, for 1996 and 1997.

Sutter, Karen M., "WTO and the Taiwan Strait: New Considerations for Business," *China Business Review*, January–February 2002a.

_____, "Business Dynamism Across the Taiwan Strait: The Implications for Cross-Strait Relations," *Asian Survey*, May/June 2002b.

"Taiwan Is Worried About Companies Moving Production Abroad," *Inside China Today*, February 4, 2002.

"Taiwan to Move 60 Percent IT Hardware Production to Mainland China in 2003," *Taiwan Economic News*, November 28, 2002.

"Task Force Formed to Check Leaks of High-Technology Secrets: Official," *Central News Agency*, March 13, 2002.

Thomas, Owen, "The Outsourcing Solution," *Business 2.0*, September 2003.

Thurow, Lester, "Microchips, Not Potato Chips," *Foreign Affairs*, Vol. 73, No. 4, 1994, pp. 189–192.

Treacy, Michael, and Fred Wiersema, *The Discipline of Market Leaders: Choose Your Customers, Narrow Your Focus, Dominate Your Market*, Boulder, Colo.: Perseus, 1995.

Tsai, Ting-I, "TSU Lawmakers Disrupt UMC's Board Meeting," *Taipei Times*, June 4, 2002.

"TSU Not Against PRC Links: Lawmaker," Central News Agency, June 16, 2002.

Tyson, Laura D'Andrea, *Who's Bashing Whom? Trade Conflict in High-Technology Industries*, Washington, D.C.: Institute for International Economics, 1992.

U.S. Bureau of Labor Statistics, *Multifactor Productivity Trends in Manufacturing, 2000*, U.S. Department of Labor, August 29, 2002.

U.S. Census Bureau, *Annual Survey of Manufacturers 1993: Statistics for Industry Groups and Industries*, U.S. Department of Commerce, Economic and Statistics Administration, MS93(AS)-1, 1995.

_____, *Annual Survey of Manufacturers: Statistics for Industry Groups and Industries: 2001*, U.S. Department of Commerce, Economic and Statistics Administration, M01(AS)-1, 2003.

U.S. Department of Defense, *Study on Impact of Foreign Sourcing of Systems* [Required by Section 831 of the National Defense Authorization Act for Fiscal Year 2001], October 2001.

U.S. Department of the Treasury, Office of International Affairs, Office of International Investment, "Exon-Florio Provisions," undated, available at www.treas.gov/offices/international-affairs/exon-florio.

U.S. General Accounting Office, *Defense Trade: Identifying Foreign Acquisitions Affecting National Security Can Be Improved*, GAO/NSIAD-00-144, June 2000.

_____, *Export Controls: Rapid Advances in China's Semiconductor Industry Underscore Need for Fundamental Policy Review*, GAO-02-620, April 2002.

USTR—*see* U.S. Trade Representative.

U.S. Trade Representative, *1999 USTR Annual Report*, March 2000.

Wang, Mark, Shari Pfleeger, David M. Adamson, Gabrielle Bloom, William Butz, Donna Fossum, Mihal Gross, Charles Kelley, Terrence Kelly, Aaron Kofner, and Helga Rippen, *Technology Transfer of Federally Funded R&D: Perspectives from a Forum*, Santa Monica, Calif.: RAND Corporation, CF-187-OSTP, 2003.

Wayne, Leslie, "Butting Heads with the Pentagon," *New York Times*, July 23, 2003.

Wessner, Charles W., ed., *Securing the Future: Regional and National Programs to Support the Semiconductor Industry, Government-Industry Partnerships*, Washington, D.C.: National Academies Press, 2003.

Yager, Loren, Susan Way-Smith, Heide Phillips Shockley, and Mary Anne Doyle, "Analysis of the 1986 Semiconductor Trade Agreement," working draft, RAND Graduate School, 1990.

Yoffie, David B., *The Global Semiconductor Industry, 1987*, Boston: Harvard Business School, 1987.

Yu-tzu, Chiu, "Technology Law to Guard Valued Assets," *Taipei Times*, April 17, 2002a.

_____, "NSC to Regulate High-Technology Labor Across the Strait," *Taipei Times*, April 18, 2002b.